P9-DCN-381

Living with Complexity

Living with Complexity

Donald A. Norman

The MIT Press
Cambridge, Massachusetts
London, England

For information about special quantity discounts, please email special_sales@mitpress.mit.edu

This book was set in Gotham by The MIT Press. Printed and bound in the United States of America.

Library of Congress Cataloging-in-Publication Data

Norman, Donald A.
Living with complexity / Donald A. Norman.
p. cm.
Includes bibliographical references and index.
ISBN 978-0-262-01486-1 (hardcover: alk. paper)
1. Technology—Social aspects. 2. Complexity (Philosophy). I. Title.
T14.5.N673 2011
601—dc22

2010012892

10 9 8 7 6 5 4 3 2

Contents

Books by Donald A. Norman

Textbooks

Memory and Attention: An Introduction to Human Information Processing (first edition, 1969; second edition 1976)

Human Information Processing (with Peter Lindsay: first edition, 1972; second edition 1977)

Scientific Monographs

Models of Human Memory (edited, 1970)

Explorations in Cognition (with David E. Rumelhart and the LNR Research Group, 1975)

Perspectives on Cognitive Science (edited, 1981)

User Centered System Design: New Perspectives on Human-Computer Interaction (edited with Steve Draper, 1986)

Trade Books

Learning and Memory, 1982

The Psychology of Everyday Things, 1988

The Design of Everyday Things, 1990 and 2002 (paperback edition of *The Psychology of Everyday Things*)

Turn Signals Are the Facial Expressions of Automobiles, 1992

Things That Make Us Smart, 1993

The Invisible Computer: Why Good Products Can Fail, the Personal Computer Is So Complex, and Information Appliances Are the Answer, 1998

Emotional Design: Why We Love (or Hate) Everyday Things, 2004

The Design of Future Things, 2007

CD-ROM

First Person: Donald A. Norman. Defending Human Attributes in the Age of the Machine, 1994. Santa Monica, CA: Vanguard

Figure 1.1
Messy desks by organized people. Some people's desks reflect the complexity of their lives. But to the person who owns the desk, everything is in its place, there is order and structure. Photograph of Al Gore by Steve Pyke. © Steve Pyke/Contour by Getty Images.

1
Living with Complexity
Why Complexity Is Necessary

The guiding motto in the life of every natural philosopher should be, Seek simplicity and distrust it.
—Alfred North Whitehead (1920/1990)

The person in figure 1.1 sits unperturbed by the apparent chaos of his desk. How does he cope with all that complexity? I've never spoken with the person in the picture, Al Gore, former Vice President of the United States and winner of the Nobel Prize for his work on the environment; but I have talked with and studied other people with similar-looking desks, and they explain that there is order and structure to the apparent complexity. It's easy to test: if I ask them for something, they know just where to go, and the item is retrieved oftentimes much faster than from someone who keeps a neat and orderly workplace. The major problem these people face is that others are continually trying to help them, and their biggest fear is that one day they will return to their office and discover someone has cleaned up all the piles and put things into their "proper" places. Do that, and the underlying order is lost: "Please don't try to clean up my desk," they beg, "because if you do, it will make it impossible for me to find anything."

My own desk is not nearly as messy as Al Gore's, but it is piled high with papers, technical and scientific magazines, and just plain "stuff," chaotic in appearance, but exhibiting an underlying structure that only I am privy to.

How do people cope with such apparent disorder? The answer lies in the phrase "underlying structure." My desk looks chaotic and incomprehensible to anyone who is unaware of the reasoning behind the many disparate piles. Once the structure is revealed and understood, the complexity fades away. So it is with our technology. Does the cockpit of a modern jet airliner (figure 1.2) look complex? To the average person, yes, but not to the pilots. To them, the instruments are all logical, sensible, and nicely organized into meaningful groups.

"Why is our technology so complex?" people continually ask me. "Why can't things be simple?" Why? Because life is complex. The airplane cockpit is not complex because the engineers and designers took some perverse pleasure in making it that way. No: it is complex because all that stuff is required to control the plane safely, navigate the airline routes with accuracy, keep to the schedule while making the flight comfortable for the passengers, and be able to cope with whatever mishap might occur en route.

I distinguish between *complexity* and *complicated*. I use the word "complexity" to describe a state of the world. The word "complicated" describes a state of mind. The dictionary definition for "complexity" suggests things with many intricate and interrelated parts, which is just how I use the term. The definition for "complicated" includes as a secondary meaning "confusing," which is what I am concerned with in my definition of that word. I use the word "complex" to describe the state of the world, the

Figure 1.2
Appropriate complexity. To the average person, the cockpit of a modern jet airplane is incredibly complicated and confusing. Not for the pilots: to them, the instruments are all logical, sensible, and nicely organized into meaningful groups. This is the flight deck of a Boeing 787.

tasks we do, and the tools we use to deal with them. I use the word "complicated" or "confused" to describe the psychological state of a person in attempting to understand, use, or interact with something in the world. Princeton University's WordNet program makes this point by suggesting that "complicated" means "puzzling complexity."

Complexity is part of the world, but it shouldn't be puzzling: we can accept it if we believe that this is the way things must be. Just as the owner of a cluttered desk sees order in its structure, we will see order and reason in complexity once we come to understand the underlying principles. But when that complexity is random and arbitrary, then we have reason to be annoyed.

Modern technology can be complex, but complexity by itself is neither good nor bad: it is confusion that is bad. Forget the complaints against complexity; instead, complain about confusion. We should complain about anything that makes us feel helpless, powerless in the face of mysterious forces that take away control and understanding.

My challenge is to explore the nature of complexity, to relish its depth, richness, and beauty at the same time that I fight against unnecessary complications, the arbitrary, capricious nature of much of our technology. Bad design has no excuse. Good design can help tame the complexity, not by making things less complex—for the complexity is required—but by managing the complexity.

The keys to coping with complexity are to be found in two aspects of understanding. First is the design of the thing itself that determines its understandability. Does it have an underlying logic, a foundation that, once mastered, makes everything fall into place? Second is our own set of abilities and skills. Have we taken the time and effort to understand and master the structure? Understandability and understanding: two critical keys to mastery.

The major issue is understanding: things we understand are no longer complicated, no longer confusing. The airplane cockpit of figure 1.2 looks complex but is understandable. It reflects the required complexity of a highly technological device, the modern commercial jet aircraft, tamed through three things: intelligent organization, excellent modularization and structure, and the training of the pilot.

Almost Everything Artificial Is Technology

Tech·nol·o·gy (noun): New stuff that doesn't work very well or that works in mysterious, unknown ways.

Technology: the application of scientific knowledge to the practical aims of human life or, as it is sometimes phrased, to the change and manipulation of the human environment.

The definition of technology as "New stuff that doesn't work very well" is mine. The more standard definition as "the application of scientific knowledge" comes from the *Encyclopaedia Britannica*. Most people seem to hold the first definition, so that commonplace things such as salt and pepper shakers, paper and pencil, or even the home telephone or radio are not considered technologies. But yes, they are indeed technologies, and as I discuss in chapter 3, even the simplest of technologies can reveal hidden complexities. Simple everyday things can be confusing if we encounter them in large numbers where each thing, though simple by itself, comes in many different varieties and forms, each requiring a different principle of operation: keeping track of which item requires what particular operation is indeed complicated and confusing. Similarly, some apparently simple things are complicated because to use them properly requires knowledge of culture and customs as well as the behavior of others.

Why has the term "technology" come to refer primarily to items that cause confusion and difficulty? Why so much difficulty with machines? The problem lies in the interaction of the complexities of technologies with the complexities of life. Difficulties arise when there are conflicts between the principles, demands, and operation of technology with the tasks that we are accustomed to doing and with the habits and styles of human behavior and social interaction in general. As our technologies have matured, especially as everyday technologies have come to combine sophisticated computer processing and worldwide communication networks, we are embarking upon complex interactions.

Machines have rules they follow. They are designed and programmed by people, mostly engineers and programmers, with logic and precision. As a result, they are often designed by technically trained people who are far more concerned about the welfare of their machines than the welfare of the people who will use them. The logic of the machines is imposed on people, human beings who do not work by the same rules of logic. As a result, we have species clash, for we are two different species, people and technology. We are created differently, we follow different laws of nature, and each of us works according to invisible principles,

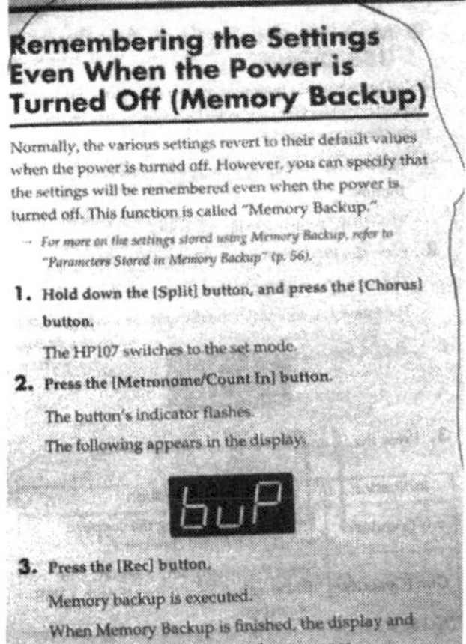

Remembering the Settings Even When the Power is Turned Off (Memory Backup)

Normally, the various settings revert to their default values when the power is turned off. However, you can specify that the settings will be remembered even when the power is turned off. This function is called "Memory Backup."

→ *For more on the settings stored using Memory Backup, refer to "Parameters Stored in Memory Backup" (p. 56).*

1. **Hold down the [Split] button, and press the [Chorus] button.**
   The HP107 switches to the set mode.
2. **Press the [Metronome/Count In] button.**
   The button's indicator flashes.
   The following appears in the display.
   buP
3. **Press the [Rec] button.**
   Memory backup is executed.
   When Memory Backup is finished, the display and buttons return to their normal appearance.

From the manual shown in the left photo: How to set the piano to remember the settings.

Hold down the [Split] button, and press the [Chorus] button

Press the [Metronome/Count In] button

(buP should appear in the display)

Press the [Rec] button.

Figure 1.3

Stupid complexity. The Roland piano is unnecessarily complicated. It's a wonderful piano, with great attention paid to the proper feel of the keys and superb rendering of the sound. But the operation of the digital controls defies comprehension. It is an expensive piano, with a really cheap display, hence the weird letters that appear. Great musicians worked on the sounds of the notes and the feel of the keyboard. Inept designers worked on the controls.

How Even Simple Things Can Become Frustrating and Complicated

Want an example of an unnecessarily complicated, frustrating device? Take my piano. The controls of the Roland piano shown in figure 1.3 are perplexing beyond belief.

The piano settings are important, for they allow the player (my wife) to make the piano sound precisely in the desired way, in our case, like a concert grand piano for playing classical music. It takes quite a while to adjust everything just right, but that's OK because there are many subtleties to be controlled, and each one seems reasonable and logical. But after all that work, it is only natural that we would want to be able to save the results so that they are always present whenever we turn on the power and start playing.

The concept of saving the settings for a device is simple enough. It is a frequent operation for any device that has numerous adjustments and settings. How are the users of this piano expected to save their settings? Here is the text from the manual (shown in figure 1.3):

1. Hold down the [Split] button, and press the [Chorus] button.
2. Press the [Metronome/Count In] button (buP should appear in the display)
3. Press the [Rec] button.

Even though my wife and I have saved the settings on numerous occasions, we can never remember the sequence and must always dig out the manual and try to do the operations. The steps are so arbitrary and unnatural that each time I have to do this, the first attempt always fails, even with the manual open in front of me.

This is an expensive piano, with a great mechanical feel to the keys and excellent acoustics mirroring the rich subtleties of the best acoustic pianos. But the company completely neglected the controls for the piano. They used a cheap, inelegant display (see the poor quality of the display letters in figure 1.3) and although there are dedicated buttons for controlling the musical sounds, there is no attention paid to other aspects of the piano setup. In other words, the piano controls were an afterthought, with no attention paid to the needs of the customer—a violent contradiction to the care and concern that went into designing the musical quality of the piano.

Usually, when I see bad design, I try to imagine what forces were involved to cause such a poor result. In this particular case, I fail. The reasons are unfathomable. Even the Owner's Manual is unintelligible. This is a design problem, and good designers can think of many elegant solutions to prevent accidental loss of the desired settings. The major cause of complicated, confusing, frustrating systems is not complexity: It is poor design.

hidden from the other, principles that harbor unspoken conventions and assumptions.

When complexity is unavoidable, when it mirrors the complexity of the world or of the tasks that are being done, then it is excusable, understandable, and learnable. But when things are complicated, when the complexity is the result of poor design with completely arbitrary steps, with no apparent reason, then the result is perplexing, confusing, and frustrating. Poor design leads to the emotional distress we have come to associate with modern technology. Good design can provide a desirable, pleasurable sense of empowerment.

There are many cries for simplicity in our lives, simplicity in the activities we pursue, the possessions we own, and especially in the technologies that we use. "Why are there so many buttons, so many controls?" people plead. "Give us fewer buttons, fewer controls, fewer features," they say. "Why can't we have a cell phone that just makes phone calls: no more, no less?" Invariably, the demand for simplicity is illustrated with wonderfully simple devices and things, simple appliances, hand tools, or household items, all with the intent of demonstrating that simplicity is indeed possible.

In attempting to reduce the frustrations caused by the complicated nature of much of today's technology, many solutions miss the point. It is no great trick to take a simple situation and devise a simple solution. The real problem is that we truly need to have complexity in our lives. We seek rich, satisfying lives, and richness goes along with complexity. Our favorite songs, stories, games, and books are rich, satisfying, and complex. We need complexity even while we crave simplicity.

The difficulty with the cry for simplicity is that many of our activities are not simple. A cell phone, for example, is expected to be able to be turned on and off (that's one control), to make and receive phone calls and then to be able to end them—that's another two controls. If we wish to dial a telephone number we need buttons for the ten digits. But even that is not enough: it's useful to store a list of frequently called people and to keep a list of who has called the phone and who has been called. We keep adding desirable actions: take photographs, play music, listen to a call with loudspeaker or earphones, and send text messages. We want to be able to do all of these things, yet we want the device to be simple. The real challenge is to tame the complexity that life requires.

Real activities are incredibly intricate with numerous components, the requirement for flexible execution, and the need for numerous alternatives. So how do we manage that complexity? Suppose a simple, small device has twenty-five buttons. Worse, suppose it has fifty. That just has to be complicated, right? Wrong.

Later, in chapters 7 and 8, I discuss the rules of design; but for now take a look at the calculators of figure 1.4. Because the many buttons are organized into logically sensible patterns, the calculator is not perceived to be particularly complex: ten number keys plus a decimal point, five arithmetic operations, a key to reverse the sign of a number, and a clear key, and four memory keys. And three buttons at the very top, dealing with the computer display. Even if the memory keys and the change sign key are novel and not understood, the calculator as a whole is sufficiently well understood that they can simply be ignored. Similarly, the scientific calculator with almost fifty keys is sufficiently well organized

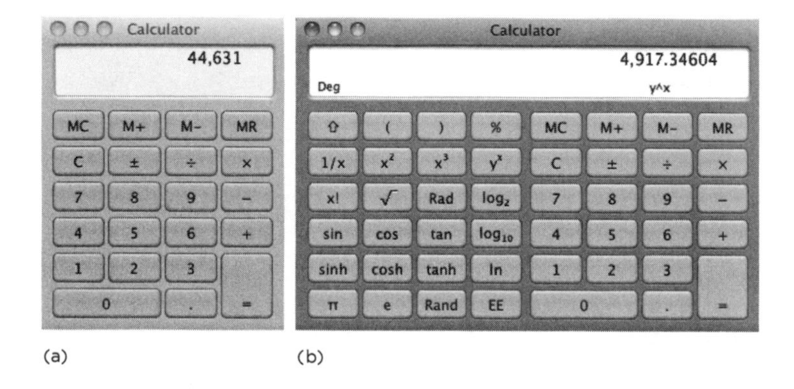

(a) (b)

Figure 1.4

Many buttons do not necessarily lead to confusion. The calculator in figure 1.4a has twenty-five buttons (including the three circular ones at the top left that control the computer window for the calculator). But because they are organized into logical groups, most people find the calculator to be simple and understandable. Similarly, the scientific calculator in figure 1.4b with forty-nine buttons (and a display) is readily understood, even by people who have no idea what the labels "sinh" "Rand," and "y^x" mean: they can simply be ignored.

that it too can be used even if not all the keys are understood. In this example, familiarity and organization are the two secrets of simplification.

Simplification is as much in the mind as it is in the device. Just imagine that the keys had been randomly arranged: the same calculator that was once easy to use then becomes quite difficult and confusing. Organizational structure makes the difference.

Complex Things Can Be Enjoyable

The world is complex. Look at the flags in figure 1.5. Does it make sense that two flags just across the street from one another should blow in opposite directions? The flags flying in opposite directions reflect the invisible complexity of nature. Note that observing the flags does not lead to irritation or annoyance so much as amusement: "Maybe we shouldn't go out today, or if we do, watch out for the wind." That is the way nature is: wind can sometimes move in mysterious, complex ways.

Some complexity is desirable. When things are too simple, they are also viewed as dull and uneventful. Psychologists have demonstrated that people prefer a middle level of complexity: too simple and we are bored, too complex and we are confused. Moreover, the ideal level of complexity is a moving target, because the more expert we become at any subject, the more complexity we prefer. This holds true whether the subject is music or art, detective stories or historical novels, hobbies or movies.

Sometimes the complexity is unexpected, but necessary, as in sports or law, where the ability of people to figure out ways around

Figure 1.5
Even nature is complex. Two flags, just across the street from one another, but blowing in opposite directions. Why? Just a typical windy day in Evanston, Illinois, just north of Chicago (also known as "the windy city"). (The photograph is authentic, taken from the window of my apartment.)

rules requires yet more rules. Today, the laws are both numerous and imprecise, so that not even our most powerful computers can master them. In sports, professional referees must sometimes huddle or call on others to determine a ruling. The American game of baseball, for example, is a relatively simple game, but the rulebook is over 200 pages long: the simple listing of baseball terms with abbreviated definitions takes thirteen pages. The same phenomenon is true of all major sports. The official rulebook for soccer from the International Football Association Board is over seventy pages long with a forty-four-page "Question and Answer" section plus a 300-page guidebook for officials. Their convenient downloadable "laws of the game" has 138 pages.

Let's take one example from baseball. The infield fly provides a good example of baseball's complexity. For readers who do not follow sports, or to whom baseball is foreign, the following text may seem inscrutably mystifying, which is precisely the point. Whatever your favorite sport, hobby, or profession, experts relish the details while others scratch their heads, amazed that human adults would spend so much time and energy on such matters. I guarantee that whatever your favorite pastime, there will be customs or rules just as arcane, just as inscrutable as that for baseball's infield fly rule.

The point of the rule is this. If the batter hits a fly ball into the infield and a member of the defending team catches it before it hits the ground, all offensive players who were running around the bases must return to the base they started from. Moreover, they are allowed to return safely. But that rule provides an interesting opportunity: Suppose the ballplayer failed to catch the ball: then it would be permissible for the player to quickly pick up the ball and trap the opposing team members off base. Defending teams soon learned that it was to their advantage to deliberately drop the ball, then pick it up quickly and tag offensive players while they were off base, eliminating them from the game. That was considered unfair, so the infield fly rule was adopted—making the infield fly ruled as caught whether it is or not, and the batter automatically out—to prevent this trick.

The rule only applies to infielders, which adds to the difficulty, for then the rule has to define who an infielder is. Why does it apply only to infielders? The purpose is to prevent infielders from deliberate manipulation of the rules to benefit their team, but who is an infielder? It turns out that even outfielders can be infielders. The rule states "*The pitcher, catcher and any outfielder who stations himself in the infield on the play shall be considered infielders for the purpose of this rule.*" What does "deliberately failing to catch" mean? Now the rule requires a commentary, which is an official part of the rulebook: "*the umpire is to rule whether the ball could ordinarily have been handled by an infielder—not by some arbitrary limitation such as the grass, or the base lines. The umpire must rule also that a ball is an infield fly, even if handled by an outfielder, if, in the umpire's judgment, the ball could have been as easily handled by an infielder.*" The full definition together with the official commentary takes around 350 words—a full page of text.

Do the complexities of baseball annoy us? Yes, of course, but they also contribute to the enjoyment of the game. Fans relish the long debates over the intricate rules. Sports journalists take pride in their detailed knowledge and in their ability to contradict the referees. The complexity of the rules adds to the sport. Moreover, there doesn't seem to be any alternative: the rules, whether of law or of a sport, are necessary to define the parameters of permissible behavior. Our behavior is complex and sometimes perverse: our rulebooks and laws mirror that complexity.

Even where complexity is not required, we sometimes seek it out. Look at the coffeemaker of figure 1.6. Is this complexity necessary? Actually, the making of coffee is a wonderful example of the trade-offs between simplicity and complexity, convenience and taste, ease versus the pleasure of drawn-out rituals.

Coffee and tea start off as simple beans or leaves, which must be dried or roasted, ground, and infused with water to produce the end result. In principle, it should be easy to make a cup of coffee or tea. Simply let the ground coffee beans or tea leaves seep in hot water for a while, then separate the grounds and tea leaves from the brew and drink. But to the coffee or tea connoisseur, the quest for the perfect taste is long-standing. What beans? What tea leaves? What temperature water and for how long? And what is the proper ratio of water to leaves or coffee?

The quest for the perfect coffee or tea maker has been around as long as the drinks themselves. Tea ceremonies are particularly complex, sometimes requiring years of study to master the intricacies. For both tea and coffee, there has been a continuing battle between those who seek convenience and those who seek perfection. Do you want the ritual of tea or coffee making, followed by luxurious enjoyment, or do you simply want to have the drink immediately, without fuss or bother? At times, we might prefer the complexity of the ceremony and the complex subtleties of the taste, while at other times, we put ease and simplicity over ceremony and ritual. The preparation of food is one case where, in the trade-off between simplicity and complexity, simplicity does not always win.

Figure 1.6
Delightful complexity. The Balancing Siphon Coffee Maker by Royal Coffee Makers. Does the coffeemaker seem overwhelmingly complex? Yes, and that is just the point: the delightful visual complexity is one of the attractions.

The quest for the perfect coffeemaker that will provide the perfect flavor with the least amount of effort is worthy of study in its own right. The options vary from the simple to the elaborate. The simplest is probably throwing cracked or ground beans into a pot of water and letting them boil a few times (three is the magic number in many countries). The most elaborate is to use expensive machines that automatically grind the beans, tamp them, heat the water, make the coffee, and dispose of the grounds. The variety of automatic coffeemakers continues to increase, from automatic drip coffee machines to today's favorite new coffee pod method. Using one prepackaged pod per cup, it provides a single cup of coffee with minimum wait and cleanup required.

An extreme case that favors complexity is the wonderful vacuum coffeemaker shown in figure 1.6. Put the water in the right-hand container, the coffee in the left. Light the fire under the right-hand container and when the water boils, the resulting air pressure forces the water into the left-hand container, where the water mixes with the coffee. The left-hand side is now heavier than the right-hand side, which causes a cover to drop over the flame, allowing the right-hand side to cool, decreasing its pressure. The coffeemaker's manual says this creates a vacuum on the right that sucks the coffee back into the container, straining the beans out as it makes its passage. I have no idea how good the coffee is, but the machine itself and the ritual are clearly the major components of the pleasure of the machine.

Why such a complex routine to make a simple cup of coffee? Rituals invariably add complexity to our lives, but in turn, they provide meaning and a sense of membership in a culture. For the coffee-lover, the intricate ritual of coffee preparation adds fun and pleasure to life. If cost and time were irrelevant we might always prefer freshly prepared food to canned and frozen food, freshly ground and brewed coffee or whole-leaf tea to instant coffee or teabags. Ultimately most of us choose which method to use depending on time factors and the importance of each event in its social context.

All cultures have rituals for food preparation and eating. When we eat we follow societal conventions: which utensils to use and for what? Who eats first, or last? Who serves or pours for whom? It is all covered by ritual. Consider these three alternatives: (A) a meal cooked by a chef who hand-chopped fresh food, sautéed the portions that needed sautéing, and spent thirty minutes preparing the food to your taste; (B) the same as (A) except that you are the chef; (C) food quickly prepared by defrosting a frozen package in a microwave oven. Which alternative would you prefer? Answer: it all depends. Life is always a complex mixture of trade-offs, in this case including time and effort, cost, taste, the pleasure of creating something, and the needs of the day.

Common Aspects of Life That Require Months of Study

One way to measure complexity is by the amount of time required to learn the item. The surprise is the number of activities that we take for granted; even activities that we like to call easy and "intuitive" are actually complex, arbitrary, and difficult to master. Some difficult things are a result of the complexity of nature and the world. Thus, the complexity of farming results from the complex intermix of the biological needs of plants, the vagaries of weather and its yearly cycle, and the care and feeding of farm animals. Food preparation is complex because of the need to transform the raw material, whether animal, vegetable, or mineral, into a form that is digestible and palatable. On top of these natural needs, people have imposed social requirements such as the elaborate rituals that accompany the preparation and consumption of food. The rules that establish what kind of behavior is proper and appropriate while dining—table manners—may be arbitrarily complex and even meaningless, but society demands that they be learned and followed. Even people who believe that they ignore the standards in fact have their own internal rituals to follow.

Society has adapted to many arbitrarily complex systems so well that adults scarcely pay any attention to their complexities and difficulties, for they have forgotten the long period of study required to master them. Two complex systems that are both complex and confusingly complicated are time specification and alphabets.

The human relationship with time has a very long history. Farming and hunting follow yearly cycles, which led to the development of calendars and timekeeping, regulated mostly by priests. The industrial revolution created factories that required multiple people to work together, both in the same location and at the same time, so the clock became an important method of controlling behavior: when to wake, eat, and pray; when to report to work; when to take a break; and when to quit. The clock, and therefore timekeeping, regulated society, even though the clock itself is an arbitrary mechanical device, not well suited to human needs.

Once upon a time, the hours of the day were specified according to human needs, with the daylight period broken up into twelve hours. Noon was the start of the sixth hour. In northern climates, the period of daylight is far longer in summer than in winter, but because the hour was defined as one-twelfth of the period between sunrise and sunset, a summer hour was much longer than a winter one. Although this method of timekeeping has been replaced with the mechanical consistency of pendulums, astronomical measurements, and atomic vibrations, the division of the day into two periods of twelve hours remains. During the French revolution at the end of the eighteenth century an attempt was made to redefine the units of time in a more sensible, decimal format. Obviously, the attempt failed.

Time is indicated by clocks with two similar-looking rotating hands, one indicating units of twelve, the other units of sixty. Many people resist the simplification of timepieces to easy-to-understand decimal displays, instead preferring the rotating analog displays that take children months to master and still give rise to interpretation errors. (The claim is that the "analog" hand allows one to make a ready estimate of time past or remaining. Try explaining that to a child struggling to learn the system.) The way we specify time is complex and confusing, but society has learned to accept it.

Alphabets form another set of arbitrary complexity. Language naturally evolved through speech and gesture. The invention of writing caused the different cultures of the world to grapple with the way by which sounds might be represented through written marks. The result is a wide variety of methods, not all of which are well matched to the sound systems of the language.

Some languages have an alphabet, with each symbol supposedly representing a sound. Some have syllabaries, where a symbol represents a syllable, usually a consonant-vowel pairing. Some languages don't have either alphabet or syllabary, just a unique ideogram for each word, so learning to read involves memorizing each character and its pronunciation, a process that continues over a lifetime: Chinese is the main such example. Japanese uses both syllabaries and ideograms, compounding the problem by having two quite different-looking syllabaries, although with the same sound patterns. Learning Japanese requires learning two syllabaries (katakana and hiragana) plus Chinese ideograms (kanji), as well as the Roman alphabet, which is used for some words and situations.

All readers of a language have had to master its writing system, but most adults forget how difficult that task was. Not only do the pronunciations for each symbol have to be mastered, but the pronunciation often varies with the context. Even the shapes of letters can be written in different ways depending on whether it is upper or lowercase, handwritten (cursive) or printed, or if it is at the start, middle, or ending of a word. Some languages use the vowel symbols only for children or people learning the language, leaving them out of adult texts. Writing systems for the different languages of the world are amazingly complex.

The tension between power and ease of learning is not easily overcome. In some languages, the relationship between the character and the sound is direct and straightforward. In others, the relationship seems bizarre and arbitrary, with English perhaps the worst example of arbitrary spelling and pronunciation.

Some languages have a carefully designed alphabet. For example, the Hangul alphabet of Korea was carefully designed in the fifteenth century by the Emperor and a committee of linguists (but continually refined even during the mid-twentieth century) to have fourteen symbols for the consonants of the language plus ten symbols for the vowels. Words are formed by arranging the characters into blocks, each comprised of three or four consonant-vowel-consonant groupings. Although the result looks a bit like a Chinese character, it is composed of alphabetic symbols, which means that the pronunciation of new words can be figured out, something that is not true with Chinese characters. Native Korean speakers perceive this to be so easy and elegant that they claim the alphabet can be mastered in fifteen minutes. One authoritative book by a linguist is entitled "You can learn the Korean alphabet in one morning." These claims are highly exaggerated.

Example: The sounds corresponding to the six English letters of the word "Hangul" are represented by the six Hangul characters "ㅎ," "ㅏ," "ㄴ," "ㄱ," "ㅡ," and "ㄹ." These characters are written in two blocks of three characters each as "한글."

I write this paragraph while I am in Daejeon, South Korea, where I have been struggling for weeks to learn Hangul, the Korean alphabet. Other non-Koreans confirm that this is how long it took them. Why is it so difficult? Yes, the alphabet is designed elegantly. But all languages have their subtleties of pronunciation and it is difficult for a writing system to capture all of the spoken sounds. English has twenty-six letters in its alphabet, but the rules of English spelling and pronunciation are incredibly complex: even native speakers make mistakes. The Korean alphabet, in addition to its ten vowels and fourteen consonants, has eleven additional vowel symbols derived from combinations of the basic vowels, five double consonants, which have their own rules, and then eleven more combined consonant rules.

In all, there are fifty-one different symbols to be learned, and although scholars insist the shapes are not arbitrary because the letter shapes are said to indicate the proper shape of the mouth and tongue in creating the sounds or phonemes, in practice this relationship is so subtle and abstract that for me at least, it plays no role in learning. Hard to learn? Complex? Yup.

Don't blame Korea for this complexity: it really does have one of the most logical and elegant of all alphabets. Blame the world. Languages have evolved over thousands of years and all have developed shortcuts, borrowed forms, special cases of grammar and pronunciation. No simple alphabet or syllabary can completely capture its inherent complexity.

This is the way of all human languages. Wonderfully expressive, wonderfully powerful. The invention of writing has enhanced our lives immensely. Writing allows knowledge, thoughts, stories, and poetry to be saved for others. It allows the dissemination of knowledge across space and time. It is the invention of artifacts such as writing that makes us smart: it is things that make us smart, things including inventions such as writing and reading. But the written marks on paper are so very different from the spoken sounds of a language that the apparent contradictions and complexities are inevitable. The spoken language is natural, learnable by anyone. The written language is arbitrary and capricious, difficult to learn, with a surprisingly large number of the world's population unable to master it.

The way we represent music has far-reaching historical roots, but that doesn't mean it is easy. For most instruments, musical notes are depicted by ovals located on staves, each staff having five horizontal lines, allowing notes to be placed below and above the staff (sometimes by adding short, temporary horizontal lines that act as extensions of the five fixed ones of the staff), with the notes being placed either on lines or between them. The lines and spaces do not allow for all the notes used in music, so other symbols, sharps (♯) and flats (♭), need to be used as well. Adding to the complexity, the notes on a staff are determined by a particular clef symbol. There are four clefs in wide use: treble, bass, alto, and tenor, so the very same set of ovals and lines means different notes in the different clefs. An oval on the bottom line of the staff has a different meaning in each clef: It is read as an E in treble clef, a G in bass clef, an F in alto clef, and a D in tenor clef. Piano players usually use two clefs, bass and treble, which means that they have to read two staves simultaneously, with different rules for each. Organ music uses a grand staff comprised of three staves, one for each hand and one for the foot pedals: usually top staff is treble clef, bottom staff is bass clef, and middle jumps back and forth. In design, when the same symbol or operation means different things depending on the context, it is called a "modal" display, and it is a well-known source of confusion and error (see figure 1.7).

The confusion in reading music is not necessary. After a little bit of tinkering, I devised a notational scheme in which every clef represents precisely one octave, so an oval always means the same thing, regardless of which clef it appears in. But then a little bit of sleuthing on the Internet revealed that I had joined a long list of innovators who have attempted to overcome the deficiencies of the scale system. The influential twentieth-century composer Arnold Schoenberg, writing almost a century ago (1924), said, "The need for a new notation, or a radical improvement of the old, is greater than it seems, and the number of ingenious minds that have tackled the problem is greater than one might think."

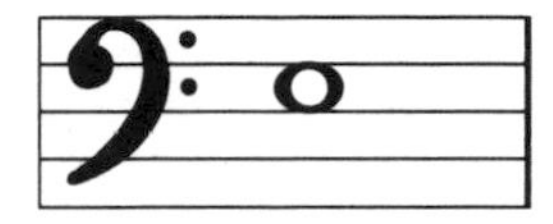

Figure 1.7
The treble and bass clefs. Illustrating the modal nature of this notation that adds to the confusion in learning: the oval of the treble clef (upper staff) indicates the musical note C, whereas the same symbol on the bass clef (lower staff) indicates an E.

I soon discovered a notational system superior to mine that eliminated the need for all those sharps and flats and other confusions brought about by keys. This was a chromatic staff, using five lines just as is used now, but assigning every note to its own position. This gets rid of the need to mark sharps, flats, or naturals, or for that matter even to indicate the key of music, except to inform the player. Thus, the bottom line of the staff represents D, but the space above that is D♯, the next line E, the space above that F, and then the next line F♯ (see figure 1.8).

Could we ever switch to this or any of the many other rational systems? Unlikely: tradition is difficult to overcome.

Musical instruments come in a wide variety of shapes, sizes, and forms. Most have a long history, sometimes thousands of

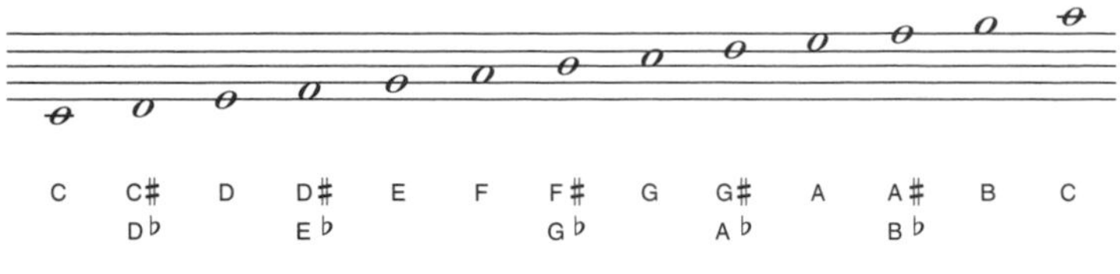

Chromatic scale on a five-line chromatic staff

Figure 1.8
Chromatic scale notation. A superior rendering of the musical staff in which sharps and flats need no longer be used, scale markings are redundant (but still useful), but most important of all, because each staff represents exactly one octave, every staff, whether higher or lower in the scale, always represents notes the same way: D, for example, is always the note on the bottom line, regardless of the octave in which it is to be played. From The Music Notation Project, <http://musicnotation.org/>.

years, and their basic structure derives in part from the accidental discoveries of early musicians, in part from the properties of the physics of vibrating strings, columns of air, membranes, and reeds. Very little attention has been paid to the ergonomics of the instruments. As a result, they often require awkward body positions, such as the contortion of the left hand required to play the violin and related stringed instruments, and sometimes exert great strain: look at the bulging cheeks of brass players, or the calluses on fingers tips of string players. Numerous musicians, especially those who use keyboards or stringed instruments, have had to end their careers because of repetitive stress injuries from playing. Many professional musicians have suffered severe hearing losses because they must endure very high sound levels for extended periods. I am convinced that if the instruments were introduced today and forced to undergo ergonomic review for health and safety, they would fail. The makers of computer keyboards, a mild device compared to many musical instruments, have been repeatedly sued in the U.S. courts for injuries to the hands and wrists.

Musical instruments are not easy to master. Even the most simple-looking can take years. The piano, for example, is relatively straightforward to understand, but incredibly difficult to master. The learning time is measured in years. Note that there are two parts to learning an instrument. One is the physical mastering of the mechanics itself: how to hold the hands, posture, and breathing. Many instruments require demanding physical exertion or special blowing techniques. Some require different rhythms in each hand simultaneously, and some require use of both hands and feet simultaneously (harp, piano, organ, percussion). But that, in many ways, is the easy part. The hard part is mastering the music, understanding the composer's and conductor's intentions, and playing in harmony with the other players. In jazz or rock music, there may be no printed score, so the performer has to improvise appropriately. These skills require a lifetime of practice.

Even the everyday activity of automobile driving, which seems easy once mastered, is complex enough that it takes weeks of instruction and then months to develop skilled performance. Remember when you first learned to drive? Everything seemed to be happening so quickly, with simultaneous actions required of each hand and foot, while watching out for cars behind, to the sides, and for objects in front, plus reading and obeying traffic signs and lights that were located at unknown places along the road: it seemed impossible. After a few years of driving, it feels so simple and easy that people eat food, put on makeup, pick up items from the floor, and do all sorts of activities while driving. The simplicity is deceiving. During normal driving, the skilled driver has lots of free time: if anything, driving is boring. But suddenly, without warning, a dangerous situation can appear. The result is that every year, tens of millions of people across the world are injured in automobile accidents.

Is driving simple or complex? Understandable or complicated? Answer: it all depends on the driver and the situation.

Do we dislike the fact that learning to read and write, play musical instruments, and drive a car are all so complex? Not really. We don't mind complexity when it seems appropriate. Yes, we truly dislike spending an hour learning some arcane, bizarre machinery. But we are willing to spend weeks or years learning other things, where the difficulties and complexity seem appropriate to the tasks: driving a car, learning the multiplication table, and rules of long division. Or learning the alphabet, and, when visiting new countries, learning their alphabets.

Think about learning to play tennis or golf, to draw and paint, or about learning a new craft. Each activity can take months to learn, years to master. I once argued that a minimum of 5,000 hours of study was required to become an expert. That judgment is today thought to be too small a number. Today, the rule of thumb among those who study skilled, expert behavior is that it takes about ten years or 10,000 hours of deliberate practice to reach world-class status.. Note that these hours do not mean merely performing or playing: they require deliberate, active practicing, often with the assistance of a teacher or coach. Expert behavior is truly difficult. These tasks have amazing complexity.

I find it interesting that we complain when a new technology requires an hour or two of study. Some people complain if only fifteen minutes of study are required. Yet we do not complain about the huge learning periods required to master the things we have grown up with, such as learning to swim, skate, or ride a bike. Reading, writing, and arithmetic, the fundamentals of education, take years to master. Should we complain about these? No, they are appropriate to the task. When new items are appropriately complex, it is reasonable that they require time and effort to master. Our complaints should be directed toward technologies and services that are unnecessarily complicated, confusing, and without apparent structure.

Figure 2.1
Chris Sugrue's *Delicate Boundaries.* The creatures wriggle around the screen, but when a hand touches the screen, they crawl off the screen and up the hand and arm. Sugrue is messing with our minds. I viewed this at her exhibition in Turin, Italy, where she won first prize. The photo comes from her Web site.

2
Simplicity Is in the Mind

After Turin, Italy, was designated "Design Capital of the World," I had to visit, in part to see their exhibition, and in part to be on a panel with Bruce Sterling, science fiction author and provocateur who was guest curator of their year-long design show. Before the panel started I meandered through the halls looking at the presentations. Sterling discovered me and declared that I had to see the exhibit by Chris Sugrue. "Why?" I asked. I had already passed by it: a computer display with moving creatures, each looking like the single-celled organisms one sees through a microscope in biology class. They wriggled across the screen in small clusters. Nice, but hardly novel. Sterling was his usual, highly persuasive self (he is a formidable debating opponent). He verbally overpowered me, dragged me back to the exhibit, grabbed my hand and placed it against the screen. The little creatures responded by moving down the screen, onto my hand, and up my arm. Huh? (See figure 2.1.)

Chris Sugrue was messing with our minds, or more accurately, with our conceptual models. When we view items on a computer screen, we know they are put there by the computer, and just as we know that images on the television screen cannot enter our living rooms, we know that crawling images of creatures on the display screen cannot wiggle onto our arms. But there they were, doing just that. Sterling was correct: it was a marvelous piece of conceptual art.

I spent some time watching other visitors as they played with the exhibit. Some people tried to brush the creatures off their arms, some tried to coax them all the way up their bodies. Nobody looked up to notice the television camera and projector that enabled the deception. A computer program was using the images from the camera to locate the person's arm and body. Then it determined when and how the images would move from the computer display to the projected image. To the viewers, the creatures were crawling off the screens onto their arms, but the anomaly was all in their minds: to the computer, this was simply a two-screen display. As I write these words, I am using a computer with two displays. I can have my writing on one screen and my notes on the other, sliding the material back and forth between the two as needed. In Chris Sugrue's artwork, the first display was the vertical screen while the second display was projected onto the horizontal surface made up of the viewer's arm: examine figure 2.1 again.

Conceptual Models

A conceptual model is the underlying belief structure held by a person about how something works. When you look at the file structure of your computer, perhaps moving a file from one folder to another, you are exploiting the conceptual model that software designers have carefully put into your head. The files and folders are fictions. There are no files or folders inside the computer. Instead, material is saved within the computer's permanent memory systems in whatever way is most convenient for the system. Most files aren't even stored in one place. Rather, they are broken up into segments, and each segment is placed wherever there is room, but with special pointers added to the file contents so that when the end of a segment is reached, the pointer tells the computer where to find the next one. In this case, the underlying complexity of the technology of storage has been replaced with the conceptual simplicity of putting files into folders, and then organizing the folders. Figure 2.2 shows the conceptual model that simplifies our understanding of computer files.

Similar fictions simplify other complexities of computer operations. For example, when you delete something from your computer, it isn't really removed. That's also a simplified fiction, part of the elaborate conceptual model underlying computer storage. Rather, the pointer to the starting segment of the information is removed, which means that under normal circumstances, the computer acts as if it isn't there. This is akin to "deleting" a book from the library by removing its entry in the catalog. If it isn't in the catalog, normal users can't ever find it, so even though the book is still on the shelves, it might as well not exist. Alternatively, a book can be "deleted" by misfiling it, moving it to an inappropriate shelf. Now it is still in the catalog, but the catalog entry does not indicate its new location. Computer experts know that "deleted" material can be retrieved by carefully examining everything in the computer's memory, ignoring the directory, ignoring

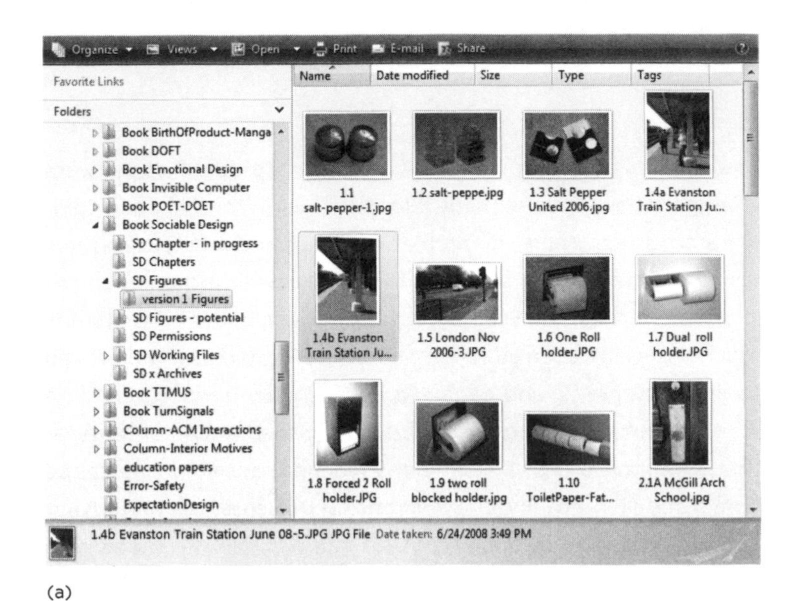
Organize
Views
Open
Print
E-mail
Share
Favorite Links
Folders
Book BirthOfProduct-Manga
Book DOFT
Book Emotional Design
Book Invisible Computer
Book POET-DOET
Book Sociable Design
SD Chapter - in progress
SD Chapters
SD Figures
version 1 Figures
SD Figures - potential
SD Permissions
SD Working Files
SD x Archives
Book TTMUS
Book TurnSignals
Column-ACM Interactions
Column-Interior Motives
education papers
Error-Safety
ExpectationDesign
Name
Date modified
Size
Type
Tags
1.1 salt-pepper-1.jpg
1.2 salt-peppe.jpg
1.3 Salt Pepper United 2006.jpg
1.4a Evanston Train Station Ju...
1.4b Evanston Train Station Ju...
1.5 London Nov 2006-3.JPG
1.6 One Roll holder.JPG
1.7 Dual roll holder.JPG
1.8 Forced 2 Roll holder.JPG
1.9 two roll blocked holder.jpg
1.10 ToiletPaper-Fat...
2.1A McGill Arch School.jpg
1.4b Evanston Train Station June 08-5.JPG JPG File Date taken: 6/24/2008 3:49 PM

(a)

9. Principles of Sociable Design.doc
Name
Date Modified
Size
Kind
9. Principles of Sociable Design.doc
Nov 22, 2009 3:39 PM
2.4 MB
Micro...ument
8 Design to the Rescue.doc
Nov 22, 2009 3:38 PM
1.5 MB
Micro...ument
7 The Power of Complexity.doc
Nov 22, 2009 3:30 PM
2.7 MB
Micro...ument
6. Clerks and Waiting Lines.doc
Nov 21, 2009 1:17 PM
1.2 MB
Micro...ument
5. Sociable Service Design.doc
Nov 21, 2009 1:07 PM
1.3 MB
Micro...ument
4 Sociable Design.doc
Nov 21, 2009 1:05 PM
872 KB
Micro...ument
3 How do we know what to do.doc
Nov 21, 2009 12:58 PM
1.1 MB
Micro...ument
don HD ▸ Users ▸ jnd ▸ Documents ▸ Writing ▸ Writing-Books ▸ Living With Complexi ▸ Sociable Design Nov 2009 ▸ 9. Principles of Sociable Design.c

(b)

the pointers. It is as if you found the lost book in the library by systematically scanning all the shelves until you found it. In the physical world, scanning thousands or millions of books is quite impractical, so librarians treat misfiled books as permanently lost. In the electronics world, even trillions of items can be scanned, which means that even when someone deliberately deletes an item, it is still there, still recoverable.

A conceptual model resides in people's minds, which is why it is also called a mental model. Conceptual models help us transform complex physical reality into workable, understandable mental concepts. The diagram of the water cycle in figure 2.3 is a good example of the power of a conceptual model to simplify our understanding of otherwise complex natural phenomena. Conceptual models are extremely important tools for organizing and understanding otherwise complex things. They enable us to understand things, learn how they work, and figure out what to do when failures occur. But just as we can watch sports games without a deep understanding of the rules, we can operate many devices without

Figure 2.2
Conceptual models: (a) and (b) show two somewhat different conceptual models for the file structures of computers. (a) shows the structure as depicted by the Microsoft operating system: the folder structure is shown in the left column and images of the files in the right window. (b) shows how Apple represents a very similar file structure with the folder structure shown at the bottom and images of the files at the top. Both fictions work well, making it relatively easy to navigate through the stored information.

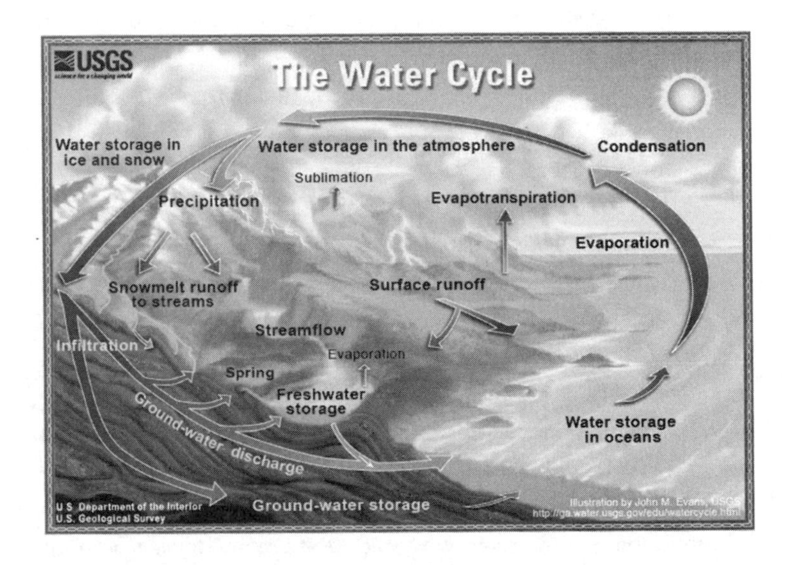

Figure 2.3
The conceptual model for the water cycle. The diagram shows a conceptual model of the way that water enters the atmosphere through evaporation, transpiration, and sublimation, and then returns through precipitation. The diagram, like most conceptual models, is a gross simplification, but nonetheless, a useful teaching model. U.S. Geological Survey drawing by John M. Evans.

understanding them, that is, without good conceptual models. We do so by following simple instructions, mimicking the actions of others, or by memorizing a standard set of actions. When some novel situation occurs, either because of the desire to do something new or because something has gone wrong, then we are in trouble: without a relevant conceptual model, we lack guidance. And when that happens we complain: Why does this have to be so complicated?

The designer's job is to provide people with appropriate conceptual models. The file structure of a computer is a good example of a concept done well. When we can see the components working, we are able to construct good conceptual models. As a result, we have built up pretty good models of mechanical products. When we deal with the electronic world where everything is invisible, we are at the mercy of the designer to provide us with hints and clues as to what is going on. And when we deal with services, even ones comprised entirely of people, we are often bewildered by the mysterious rules and regulation of bureaucracy, to say nothing of the mysterious people in the back rooms who control what the people we deal with are allowed to know and talk about.

We humans are always seeking explanations, always seeking to understand what is happening. These explanations come from our conceptual models, sometimes newly created while we are trying to understand our experiences. They apply to our views of how other people react, to the explanations we give others about our own actions. They most definitely apply to how we feel when interacting with products and dealing with services. Nameless, faceless bureaucracies can ruin a day, while friendly interactions with pleasant merchants, salespeople, and service representatives can redeem it.

Conceptual models apply to almost everything we do in life. The more complex the activity, the more important the conceptual model. Whenever a system is well understood, the average person manages quite well. By all accounts, driving is a difficult, complex activity. In the modern automobile, much of the technology is completely inaccessible to the average driver. More and more of the automobile's operations are controlled by computer chips located throughout the vehicle, networked together, responding to multiple sensors and controlling many actuators and functions. We manage to drive quite successfully because the conceptual model is readily understood. Note too that driving is not an automatic activity: most drivers are taught by individual tutors supplemented by classroom activities, books, videos, and tests. Although driving is a complex activity requiring skilled control of a fast-moving vehicle, understanding a large body of cultural norms and legal requirements, often conducted while conversing with other people, listening to music, and so on, it can be mastered.

What makes something simple or complex? It's not the number of dials or controls or how many features it has: It is whether the person using the device has a good conceptual model of how it operates.

Why Can't Everything Be as Simple as a Planishing Hammer?

There is always an easy solution to every problem—neat, plausible and wrong.
—H. L. Mencken (1917)

Our lives are too complex. Products are far too complex. It is a worldwide problem. The solution? It's obvious, and plausible: make things simpler. "Why can't products be simpler?" cry the reviewers in the newspapers, magazines, and TV shows. "We want simplicity," cry the people befuddled by all the features of their latest whatever. Do they really mean it? No, for whenever journalists review simple products, they complain that the devices lack what they consider to be "critical" features. What do people mean when they ask for simplicity? They want the simplicity of one-button operation, but with all of their favorite features. It simply is not possible.

When I published my first essays on simplicity, they were met with great skepticism. After all, I myself had decried the contagious disease that I labeled "featuritis," whereby each new version of a product adds more and more features to its set of capabilities. Each new competitor feels compelled to match the set and to add even more to be able to advertise enhanced capabilities. Over time, products became more and more complex. Featuritis was a deadly disease, difficult to avoid. There are no known vaccinations, no known cures except to start over again. So why am I suddenly arguing against simplicity?

One of my correspondents told me about the planishing hammer, a tool used by silversmiths. "Show me a silversmith's planishing hammer with added complexity," he said, "and I'll show you an unsold tool. Same for many hand tools."

At first reading, it would seem that my correspondent has a point. Craftspeople, those who live by their tools, have simple, well-designed ones. And this is not just true of silversmithing but of any specialized activity. Consider woodworking, blacksmithing, gardening, camping, hiking, and mountain climbing. The tools of professional carpenters are often simpler than the complex, multipurpose tools sold to the do-it-yourself hobbyist. Why is it that the tools of the professional crafts workers always seem simple yet everyday consumer products are so complicated?

But wait a minute. Are those tools so simple? Let's go back to the planishing hammer. I realized I had never even heard of this tool. I looked it up in a dictionary: a specialized hammer made for toughening and smoothing a metal surface. Here is what Wikipedia has to say about it:

> After a piece of metal has been roughly formed by techniques such as sinking or raising, the surface will have irregular indentations and bumps. To remove these imperfections, the piece is hammered between a flat or slightly curved hammer and a special forming object known as a planishing stake. Using repeated, relatively soft blows, the piece is smoothed toward the curvature of the stake. . . . Since planishing hammers are generally in contact with the outside surface of the piece, they have rounded edges and are kept polished to avoid marring the work.

Hmm, that doesn't sound so simple. The hammer is simple, but the usage is rather advanced: "mystifying" is the word I would use. In fact, there are books explaining how to use the hammer: that doesn't sound simple to me. The apparent simplicity of the planishing hammer is like the apparent simplicity of a unicycle or a surfboard or a pair of skis. All are very simple devices, easy to understand at a single glance, but all take years of practice to master. To call these things simple is misleading.

Consider a tool that everyone agrees is complex: a computer application for photographic editing. The professional-level programs have a large number of menu items, many labeled with exotic, specialized names. It contains so many brushes, pens, layering tools, and such an array of possible tools and operations that bookstores devote many feet of space to specialized books explaining how to use the programs. There are even schools that hold year-long courses on photo editing. Now that is complex, complicated, and to the novice, confusing.

Now consider that simple planishing tool. Is it fair to compare the simplicity of the silversmith's planishing hammer, figure 2.4a, with the complexity of the choices confronting a user of the photo-editing tool? No, it isn't fair at all. We need to compare the editing program with the crowded bench of tools used by the accomplished silversmith: figure 2.4c. Moreover, the silversmith has a wide choice of hammers: see figure 2.4b. Now the planishing hammer doesn't look quite so simple: the complexity of choices facing the silversmith seems even more daunting than the choices faced by the photo editor. We need to compare the years of training of the skilled photographer with the years of training of the skilled silversmith. The planishing hammer needs to be compared with one of the menu choices from the photographic editing application.

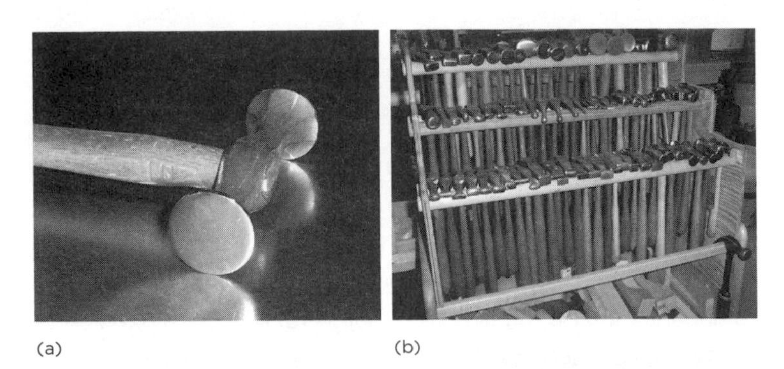

(a) (b)

(c)

Figure 2.4
Jeffrey Herman's silversmithing workbench and tools. (a) shows his planishing hammer, (b) his collection of hammers, and (c) his workbench. Sure, the planishing hammer is simple and elegant, but when it is embedded among all the other tools available to the silversmith, how do you know when to use it? Figures provided by Jeffrey Herman, President of the Society of American Silversmiths, <http://www.hermansilver.com>.

With the photographic editing application, any given menu choice is actually fairly simple: the trick is knowing which choice to make, and then to possess the delicate hand–eye coordination needed to modify the photograph in precisely the manner required. But isn't it the same for the silversmith? A novice silversmith is just as bewildered by the array of tools. When we compare the complex array of tools that skilled craftspeople select for their tasks, we see that the real complexity does not lie in the tools, but in the task. Skilled workers have an array of tools, each carefully matched to a particular task requirement. It can take a long time to learn which tool goes with which task, and years to master the tools. Thus, even when I know what a planishing hammer is for, I am certain that my use would be more likely to damage the object than to improve it. To me, ignorant of silversmithing, the images of figure 2.4 are complicated and bewildering. To an accomplished silversmith, they are probably familiar and simple.

The simplistic approach would be to ask the silversmith to use a single hammer for everything. In many cases, life is simpler by having a few complex, multipurpose tools rather than a lot of special-purpose ones. If I am traveling, I prefer a Swiss army knife, but I would never use it at home where I have a vast array of specialized knives, scissors, and screwdrivers.

Whether something is complicated is in the mind of the beholder. Even my word processing program (Microsoft Word), often held up as an extreme example of complexity gone amok, simplifies my life. To find the definition of the word "planishing," all I had to do was point at it, click on the right-mouse button, select the choice "Look Up," and be presented with the definition. The option was smoothly handled by the menu called up by the right button click: I was presented with a choice of actions. The menu itself has an underlying complexity, for it is context sensitive: the menu that appears when I push the right-mouse button depends on what task I am performing at the moment. This neatly illustrates the paradox of the quest for simplicity: to make our lives easier, we need more powerful, more complex tools.

Complexity can be tamed, but it requires considerable effort to do it well. Decreasing the number of buttons and displays is not the solution. The solution is to understand the total system, to design it in a way that allows all the pieces to fit nicely together, so that initial learning as well as usage are both optimal. Years ago, Larry Tesler, then a vice president of Apple, argued that the total complexity of a system is a constant: as you make the person's interaction simpler, the hidden complexity behind the scenes increases. Make one part of the system simpler, said Tesler, and the rest of the system gets more complex. This principle is known today as "Tesler's law of the conservation of complexity." Tesler described it as a tradeoff: making things easier for the user means making it more difficult for the designer or engineer.

> Every application has an inherent amount of irreducible complexity. The only question is who will have to deal with it, the user or the developer (programmer or engineer). (Tesler and Saffer 2007)

With technology, simplifications at the level of usage invariably result in added complexity of the underlying mechanism. Consider the automatic transmission in the automobile, a complex mixture of mechanical gears, hydraulic fluids, electronic controls, and sensors. Fewer complications for the driver are accompanied by more complexity in the underlying machinery. Simplicity must always be measured from a point of view. What is simple on the surface can be incredibly complex inside; what is simple inside can result in an incredibly complex surface. So from whose point of view do we measure simplicity?

Why Fewer Buttons May Make Something Harder to Operate

A remote control for a television set with only a few buttons may look simpler than one with one hundred, but not if the simpler-looking device requires arbitrary sequences of button pushes to get the desired result. The complex-looking device has one button per function, so even a novice can use it by searching for the appropriate label and pushing its button. Many designers equate simplicity with a simple-looking appearance, but simpler-looking does not always translate to simplicity of use.

Perceived simplicity is not at all the same as simplicity of usage: operational simplicity. Perceived simplicity decreases with the number of visible controls and displays. Increase the number of visible alternatives and the perceived simplicity drops. The problem is that operational simplicity can be dramatically improved by adding more controls and displays. The very things that make something easier to learn and to use can also make it be perceived as more difficult: This paradox is a challenge to the designer.

Simplicity is a mental state, highly coupled with understanding. Something is perceived as simple when its actions, options, and appearance match the person's conceptual model. As a result, operational simplicity is optimized when every possible action has a single dedicated control, even though this adds to the number of controls, thereby increasing perceived simplicity. With dedicated controls, it is easy to understand what each one does. Simplicity decreases when the design makes it difficult to know what is happening or when controls have multiple meanings depending on context.

In the early days of the graphical user interface, there were many debates about the appropriate number of buttons to put on a mouse. Apple decided that perceived simplicity should dominate, so they used only a single button. I once tried to find out why Apple had selected a single button. People involved in that decision told me that for beginning users of computers, having several buttons on the mouse was confusing: "When we tried two buttons, people could never remember which was which. It was even worse with three." But those early studies also showed that experienced users always preferred multiple buttons. Apple decided that they should favor the inexperienced users, so they adopted the mouse with a single button as their standard.

Was Apple correct? I suspect that their decision was wise, at that particular time. To understand this, you have to realize that the general public had never experienced a computer that used a mouse. The two previous attempts to sell mouse-driven computers to the public had failed (the Xerox Star and the Apple Lisa), so Apple was prudently cautious. The reality is, however, that a single button is simply not enough. Apple always had a second button, it simply wasn't on the mouse: It was the "Apple" key on the keyboard. Many mouse operations required the use of the Apple key.

Which is simpler: a mouse with two buttons, left and right, or a mouse with one button on the mouse, the other on the keyboard? Surprisingly, in terms of ease of use, I believe that the combination of button on mouse and keyboard is easier than the left-right buttons on the mouse. Why? Psychological studies show that left-right confusions are extremely common. It is easy for everyone to learn the difference between top and bottom, but left and right cause great difficulties for children that persist until adulthood for many people. In the annals of human error, left-right confusions show up frequently, top-bottom ones almost never. Moving one of the buttons to the keyboard gives it a unique status, making it extremely unlikely that it would ever be confused with the button on the mouse. This means that learning would be enhanced with the separate locations, even if usability is somewhat impaired. With enough practice, people can learn the left-right distinction and perform well. In the early days of mouse-driven computers, it was essential to get novice users comfortable as quickly as possible.

When I was at Apple, I tried to get them to switch to a two-button mouse. I suggested that by that time, everyone was familiar with the mouse, so the earlier objections would no longer apply. Microsoft had proved the virtue of a two-button mouse by using the right button to provide contextual information: menus and help. But the use of a single button was an important branding symbol for Apple and my efforts to change this went nowhere. Today, however, Apple uses a multiple-button mouse.

Is the single-button mouse less complicated than the multiple-button mouse? Once again, it all depends on whose point of view we are taking.

Chapter 2

Misunderstandings about Complexity

All other things being equal, the simplest solution is to be preferred.
—William of Ockham, fourteenth century

Things should be made as simple as possible, but not simpler.
—Attributed to Albert Einstein, twentieth century

Simplicity by itself is not necessarily virtuous. Two of the most famous statements about complexity in science are Ockham's Razor and Einstein's saying. Both sayings are simplifications of the actual words used. Ockham's Razor comes from the writing of William of Ockham, who in the fourteenth century argued that "all other things being equal, the simplest solution is to be preferred." (The actual language is "entities should not be multiplied beyond necessity.") Albert Einstein, the famous twentieth-century physicist, is reported to have said that "things should be made as simple as possible, but not simpler," although once again, what he actually said was something like "The grand aim of all science . . . is to cover the greatest possible number of empirical facts by logical deductions from the smallest possible number of hypotheses or axioms."

Usually, these are both interpreted in the same way: the simpler, the better. What is forgotten is that Ockham said that this was to be the rule only if "all other things being equal." For Einstein, the critical phrase is "but not simpler." Many who seek simplicity forget these two restrictions.

Ockham's Razor is meant to apply to the choice between two scientific theories, each of which explains precisely the same phenomena, but one is more complex than the other (that is, it contains more assumptions or has a more complex formulation). In fact, it is seldom the case that this condition exists: two competing theories almost invariably cover different phenomena, even if there is substantial overlap. In the case of Einstein's dictum, the same issues arise with the phrase "as simple as possible." Which is to be preferred, a simple theory that explains only a few things, or a more complex theory that explains much more?

Simplicity Does Not Mean Fewer Features

Complexity is an inescapable part of the world we live in. But complexity need not turn into complicated confusion. Complexity can be tamed through proper design. Why the cry for simplicity? It is an honest reaction to the confusion and complications of life; but although the intention is admirable, the proposed solution is mistaken.

Everyone wants simplicity, but that request misses the point. Simplicity is *not* the goal. We do not wish to give up the power and flexibility of our technologies. My single-button garage door opener may be simple, but it hardly does anything. If my cell phone only had one button it certainly would be simple, but all I could do would be turn it on or off: I wouldn't be able to make a phone call. Is the piano too complex because it has eighty-eight keys and three pedals? Surely no piece of music uses all of those keys. So should we simplify it? The cry for simplicity misses the point.

If we watch potential customers in stores, we see that simplicity does not win: people really want features. How does this correspond with their stated preference for simplicity? The conflict is easy to understand. People want their devices to be powerful, capable of all the tasks they demand of them. At the same time, they want them to be easy to use. As a result, even as people buy the devices with extra features, they cry out for simplicity. Features versus simplicity: are these two really in serious conflict?

There is an implicit assumption at work here:

More Features → Increased Capability
Increased Simplicity → Increased Usability

These two statements translate into simple logic: Everyone wants more capability, so therefore they want more features. Everyone wants ease of use, so therefore they want simplicity.

Alas, this simple logic is false logic, false because it follows the implications backward. Suppose I said:

A sunny day → it won't rain.

Does this mean that if it doesn't rain the day is sunny? Of course not. The arrow goes left to right: this says nothing about the right-to-left direction. So extra capability does not require more features. Similarly, usability does not require simplicity.

I conclude that the entire argument between features and simplicity is misguided. People might very well desire more capability and more ease of use, but we should not equate these with more features and more simplicity. What people want are usable devices, which translates into understandable ones. The whole point of human-centered design is to tame complexity, to turn what would appear to be a complicated tool into one that fits the task, one that is understandable, usable, enjoyable.

Why the Commonly Assumed Trade-off between Simplicity and Complexity Is Wrong

Some people, hearing that I was writing about complexity, suggested that this is simply the well-known trade-off between simplicity and complexity. No: the trade-off is wrong, for it is based on a false assumption. As I have shown, simplicity is not the opposite of complexity: complexity is a fact of the world, whereas simplicity is in the mind. The trade-off assumes two things: first, that simplicity is the goal; second, that one must give up something in order to gain the desired simplicity.

The trade-off is wrong because the real goal is understanding, usability, and, of course, whatever functions are desired. A trade-off assumes what is called a "zero-sum game": to get more simplicity one must get rid of complexity. But there is no need to trade essential complexity for the understanding that is just as essential. Complexity is often necessary. The design challenge is to manage complexity so that it isn't complicated.

People Like Features

Whenever I visit a country new to me, one of my favorite pastimes is to visit the local stores and markets where the inhabitants live and shop; what better way to get to know the local culture?

Foods differ, clothes differ, and in the past, appliances differed, whether kitchen utensils, gardening tools, or shop tools.

On my first few trips to South Korea, I asked my hosts to take me through the shopping markets of the city and especially to their department stores. In the department stores I found the traditional "white goods," refrigerators and washing machines. The stores obviously stocked products of the Korean companies LG and Samsung, but also merchandise from General Electric, Braun, and Philips. The Korean products seemed more complex than the non-Korean ones, even though the specifications and prices were essentially identical. "Why?" I asked the two design students who were my guides. "Because Koreans like things to look complex," they responded, "it's a symbol: complexity indicates status."

I've found the same phenomenon in the United States and Europe. Expensive, stainless steel stoves in kitchens, even where the residents seldom cook. Fancy washing machines, even though their owners apologetically admit that they cannot figure out the settings.

Appliances have increased in complexity, especially ones that used to be quite simple: for example, toasters, refrigerators, and coffeemakers, all of which have multiple control dials, multiple LCD displays, and numerous options.

Once upon a time, a toaster had one knob to control the degree of toasting—that was all. A simple lever lowered the bread and started the operation. Toasters were inexpensive. But in today's stores, toasters are quite expensive, often adorned with the names of famous designers or design firms and boasting complex controls, motors to lower the untoasted bread and to lift it when finished, and LCD panels with cryptic icons, graphs, and numbers. Simplicity?

Examine the modern automobile. Complexity again. I'm old enough to remember when the steering wheel was just for steering, the rearview mirror just a mirror. Today, steering wheels are complex control structures with multiple buttons and controls, including volume controls for music and telephone plus multiple stalks to control turn signals, cruise control, headlights, and windshield wipers. Rearview mirrors also now have multiple controls and displays.

Why do people buy an expensive, complicated toaster when a simpler, less-expensive toaster would work just as well? Why all the buttons and controls on steering wheels and rearview mirrors? Because these are the features that people believe they want. They make a difference at the time of sale, which is when such features matter most. Why do we deliberately build things that confuse the people who use them? Answer: because the people want the features. Because the so-called demand for simplicity is a myth whose time has passed, if it ever existed.

Make it simple, and people won't buy. Given a choice, they will take the item that does more. Features win over simplicity, even when people realize that features mean more complexity. You do it too, I'll bet. Haven't you ever compared two products side by side, feature by feature, and preferred the one that did more? Shame on you! You are behaving, well, like a normal person. The complex, expensive toaster? It sells well.

What really puzzles me, though, is that when a manufacturer figures out how to automate an otherwise mysterious operation, I would expect the resulting device to be simpler. Nope. Here is an example: Siemens developed a washing machine that, to quote its Web site, "is equipped with smart sensors that recognize how much laundry is in the drum, what kind of textiles the laundry load comprises, and if it is heavily or lightly soiled. Users only have to choose one of two program settings: hot and colored wash, or easy-to-clean fabrics. The machine takes care of the rest."

Hurrah, now the entire wash can be automatic, so only two controls are needed: one to choose between "hot and colored wash" and "easy-to-clean fabrics," the other to start the machine. Nope. This washer had even more controls and buttons than the nonautomatic one. "Why even more controls," I asked a friend who works at Siemens, "when you could make this machine with only one or two?"

"Are you one of those people who want to give up control, who think less is better?" asked my friend. "Don't you want to be in control?" Strange reply. Why the automation if it isn't to be trusted? And yes, actually, I am one of those bizarre people who think that less is better.

It appears that marketing won the day. And I suspect marketing was right. Would you pay more money for a washing machine with fewer controls? In the abstract, maybe. At the store, probably not. Marketing rules—as it should, for a company that ignores marketing is a company soon out of business. Marketing experts know that feature lists influence purchase decisions, even if the buyers realize they will probably never use most of the features.

Notice the wording: "pay more money for a washing machine with fewer controls." An early version of this material was published in *Interactions*, the magazine for professionals in the field of human-computer interaction. The editor flagged the sentence as an error: "Didn't you mean 'less money'?" she asked. Her question makes my point precisely. If a company spent more money to design and build an appliance that worked so well, so automatically, that it only needed an on-off switch, people would reject it. "Why does the simple one cost more than the more powerful, complex one?" they would complain. "What is that company thinking? I'll buy the cheaper one with all those extra features—after all, it's better, right? And I'll save money, too." Yes, we want simplicity, but we don't want to give up any of those cool features.

What is one of the most complex objects in our lives? Human beings. The human body—and especially the human brain—is incredibly complex. The brain evolved in fits and starts, leaving remnants of the past, reusing old stuff for entirely new purposes. I often complain of the dreadful disease of "featuritis" that afflicts the modern digital appliance. But when it comes to features, biological structures win. All biological structures are filled with features and adjustments. It takes us years to learn to control our own bodies. It takes us years to learn to use even the most basic types of products, for example, pencils and eating utensils, whether they be knives and forks or chopsticks. We soon forget how many years of our childhood were spent learning fundamental skills. Complexity is unavoidable.

Complex Things Can Be Understandable; Simple Things Can Be Confusing

The street scene shown in figure 2.5 is complex, but understandable, although not necessarily at first encounter. The panel of light switches in figure 2.6 is not complex, but it is horribly confusing. Simplicity is not the answer.

Some cultures seek simple, clean appearances. Western designers often prefer clean, spare designs with lots of space between

福生祥香莊

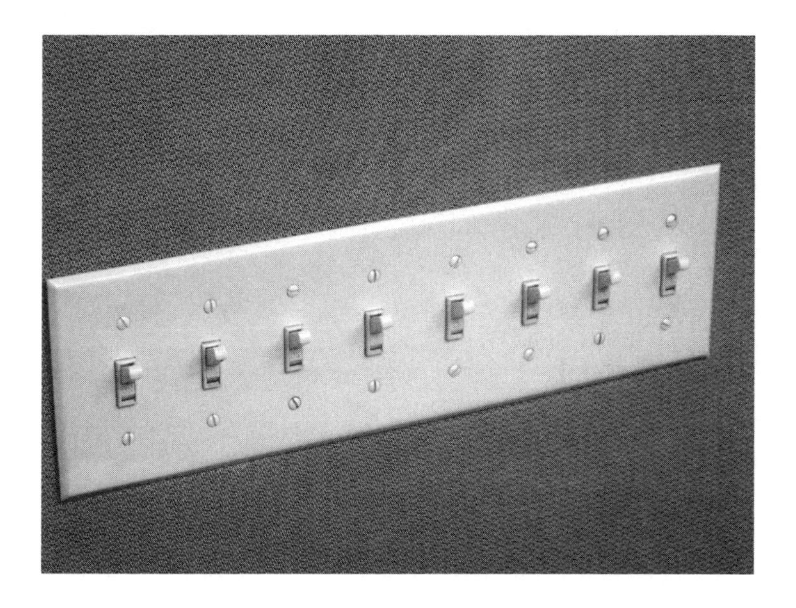

Figure 2.5
Complex, yet understandable. Cities are complex, yet understandable, although not necessarily upon first exposure. I took this picture in Hong Kong, but similar or even more complex scenes can be found in any large city across the world.

Figure 2.6
Here is a simple bank of light switches. Simple, yet confusing. Who can remember what each switch does?

elements: "white space," they call it. Eastern designs seem crowded and chaotic in contrast, but this is what is preferred. Cities in Asia bubble with activity, multiple electronic signs crowding the air, street vendors hawking their wares, political speeches attacking the senses via powerful loudspeakers mounted at street corners or on motor vehicles driving slowly along busy streets. Signs are bustling with information, every bit of space packed with images. Japan may be known for its elegant art and gardens, with simple lines and elements: raked sand, carefully placed rocks and pruned trees. But leave the serenity of the private gardens and the street life overwhelms the senses. Electronic signs, software, and Internet sites seem to fill up every available space with conflicting colors and moving images. In many ways, the most preferred designs in Asia violate the canons of tasteful design in the West.

The differences in visual preferences among cultures need to be respected by designers. Sparse, clean designs have an aesthetic appeal, but as we have already seen, they may be more difficult to use than crowded complex-looking designs, where many different alternatives and options are always on display. Apparent complexity varies with culture as well as with experience. Psychologists have long studied the nature of people's aesthetic preferences. One fundamental principle is that there is a preferred range of complexity: things that are too simple are boring, shallow. Things that are too complex are confusing, upsetting. People prefer an intermediate level of complexity. Moreover, that preferred level varies with knowledge and experience. Complex things can be easy to use; simple things can be complicated. Sometimes we prefer the complex, sometimes the simple. Taming technology is a psychological task, not a physical one.

Figure 3.1
One simple thing is simple. But many simple things, each with different roles, can be complex. Wherever there is a need for labels, it is a sign of difficulty. Any given door lock is simple; the complexity results from the wide variety we encounter daily. It isn't easy to remember how to work each door, because each is different. How do we cope? We put information on the items: words, dots, arrows, and pictures—all helping us cope.

3
How Simple Things Can Complicate Our Lives

Complex things do not have to be confusing. Similarly, confusing things do not need to be complex. Even simple things can lead to bewilderment: doors, light switches, stovetops. It isn't that any of these simple things is difficult to understand, it is that each seems to have its own way of working, so that when you first encounter yet another new example, it can be annoying and frustrating. A single example of a simple thing is just that: simple. But when there are many simple things, each with its own rules of operation, the result is complexity.

Consider the locks and keys of figure 3.1. Why should the lock controls shown in the figure be complex? Just rotate them to lock, rotate them the other way to unlock. Similarly for the keys: insert and rotate. What is complicated? If there were only one knob in the world and only one lock with a key, they would indeed be simple to use. The problems occur when each of us must deal with numerous knobs and keys. With the knobs, one rotates counterclockwise to lock, the other clockwise: how can I remember which is which? Answer: without some visual indicator, I cannot keep track. Notice that in all four parts of figure 3.1, people have added indicators to help. Any time you see signs or labels added to a device, it is an indication of bad design: a simple lock should not require instructions. Worse, the users of the devices should not have to add their own signs. Even the simplest operation can become complicated when it is one of many operating in different, arbitrary ways.

Everyday life is often complex, not because any given activity is complex, but because there are so many apparently simple activities, each with its own set of idiosyncratic requirements. Take a large number of simple actions and add them up, and the overall result can be complex and confusing: the whole is greater than the sum of its parts.

The litany of actions and decisions probably sounds trivial and unimportant, and indeed, it is. But add these simple things to the host of other simple decisions that occur during the day and the result can be overwhelming. Does the key in this lock go clockwise or counterclockwise? Do I fill the gasoline tank in this automobile on the right or on the left? Which button on the remote controls the sound, which the channel? These tiny details create a constant undercurrent of stress.

Consider the frequent demands for passwords. Most people prefer easy-to-remember passwords, such as their name, their spouse's name, or the name of their pet. When security experts study the passwords that people select on their own, they are appalled. One of the most common passwords is simply the word "password," modified to "password1" when a number is required. Other common passwords include "123456," "jesus," and "love." Security experts are horrified at this practice because these passwords are very easy for an intruder to guess or discover: just a few minutes on a social network and the bad guys have uncovered lots of personal information about people, the same information often used as passwords. As a result, the experts have added requirements to passwords: they have to be long, they must contain letters and numbers, lowercase and uppercase, and sometimes other symbols as well. They must be changed frequently, and any password used previously cannot be used again. Simple words of the language are not permitted. All these requirements are well intended and sensible. But they make the simple task of selecting and remembering a password into a complex activity. Moreover, because we all have numerous passwords, the complexity grows even greater.

When security experts insist that we all follow complex rules for the generation of passwords, often requiring that they be changed every few months, they do indeed make it more difficult for thieves, criminals, and mischief-makers to discover them, but they also make it impossible for us to remember our own passwords. Note the similarity of the password problem with the complexity of locks. If only there were one lock or one password, we could easily cope with any requirement. As the numbers get large, life becomes complicated. I've tried to explain this to security experts, usually without success. I've tried to explain that many of the requirements they impose on us to increase security actually diminish it. My own university seems to have decided that I am a crackpot, best to be ignored.

How do people cope? They write down their passwords and paste them in some convenient but hidden spot, such as underneath the keyboard. Actually, a surprising number of people just paste them to the front of their computer screen. My wife and I record all the different passwords and secret codes we need for all of our Internet activities in a special computer file: that file is now nineteen pages long in tiny font: 5,000 words of text! Yes, we encrypt the file so that even if someone gained access it couldn't be read, but this adds one more password to our memory load: the password for the file of passwords. Some companies prosper because of these problems, selling special programs to help manage the large number of passwords. A number of commonly used programs try to simplify life by automatically filling in forms with names, address and credit card information, and, of course, login names and passwords. This does indeed simplify life for the authorized user of the computer, but also for any thief who manages to get access.

Many people solve the multiple password problem by using the same password for different activities, which violates all the security rules, of course. (Even many security professionals admit, in private, that this is what they do.) Many people cope by making up easy-to-remember passwords, refusing to change them as frequently as ordered, writing them down and sticking them in places where they are easy to find: people store house keys under doormats, passwords on notes taped under keyboards or even, as in figures 3.2a and 3.2b, on the monitor itself. We put the information out in the world, because our heads are too cluttered.

Put Information in the World

When the large amount of simple, trivial information that has to be remembered ends up making even the simplest of tasks complex

(a)

(b)

(c)

Figure 3.2

Coping with passwords. We write them down and place them where we can find them. The paper pasted between the top of the keyboard and the bottom of the screen in (a) says "User Name askaggs Password 960chdAS." The paper posted on the screen in the photograph of (b) says "THE PASSWORD IS CHAIR" (this was at a furniture maker). Note the wastebasket propping up a security door in (c). The card reader on the wall is supposed to provide security, allowing access only to people with cards, but it interferes with workers who need to run back and forth to other offices outside the secure area. More onerous security requirements can lead to less secure situations.

and confusing, how can we cope? The answer is simple: put the required information in the world itself. Of course, this solution doesn't help us deal with passwords because when we put them in the world (as in figures 3.2a and 3.2b), they are no longer secret, defeating the purpose. But many of the things we have to remember are not secret: making them available in the world benefits everyone.

Look at the airplane parked at the gate in figure 3.3. How does the airplane know just where to stop so that it is lined up properly with the walkway? By signs painted on the ground showing where the nose wheel of each different kind of plane should stop to ensure that the airplane doors are in the correct position: see figure 3.3. (The pilots can't see the wheels underneath them, so airport personnel use hand signals to direct the pilots, stopping the plane when the nose wheels have reached the right spot.) The same issue shows up for all the equipment airlines need to ready their airplanes: how can they figure out where to keep them so that they are convenient and ready for use, but still out of the way? By painted lines and marks, as in figure 3.3. The principle is widely used in manufacturing plants. If you ever get a chance to visit a well-organized factory, note the use of lines, markers, and physical barriers as memory and organizational aids: "Visual Workplace—Visual Thinking" is how one book describes the power of this kind of information.

If factories and airlines can do it, so can you. The secret to coping is to take charge: stick up little reminders, warnings, pictures. Put reminders where you can find them when needed: lines, marks, sticky labels, and instructions. This dramatically simplifies what must be remembered, making it unnecessary to remember

Figure 3.3
Airports use painted marks to indicate where things go: marks to show where trucks should park, where airplanes should stop. If they can do it, so can you.

the idiosyncrasies of each case. When the information is in the world, then it doesn't clutter up the mind, but it is there when you need it.

Years ago one of my graduate students, Hank Strub (now a seasoned professional), taught me the power of sticky dots. He recommended buying packages of those little round, colorful, sticky circles and placing them wherever one had to remember some simple action. "Color coding labels," one package calls itself. "Little green dots" is what I call them, regardless of their actual color. Stick them by the controls you need to remember, stick them in the direction you need to turn the key, stick them by the lock to let you remember which way means "locked," which "unlocked. Today I make sure I always have a package of them nearby. Any color will do; all that matters is that they be visible. They are on my office locks, on all the dials of my audio equipment, on electric sockets so I can remember which ones are controlled by wall switches, and even on the dashboards of my cars so I can remember which side of the car the gasoline filler is on. One of my reminders is shown in figure 3.1: the dot indicates which way the knob should be to lock the door. Now, every night when I check the doors to make sure they are locked, I simply walk by each and see if the knobs line up with the dots.

You do need to take some care, however. Sometimes the solution can be as puzzling as the initial problem. In the case of figure 3.1, for example, how does someone know that the marked location means locked rather than unlocked? There are basically three solutions to this one. First, just learn the rule. Hmm, not a good solution unless there is only one rule, universally followed. Second, use color-coding, so that red would be shut and green open. This entails two problems: whether the color code is universally known, and color blindness—roughly 10 percent of human males are red-green colorblind, yet red and green are the most common colors used to mark things like on-off, stop-go or locked-unlocked. Finally, we could use the linguistic convention of "markedness." In this approach, the normal state is never marked: it is the abnormal state that gets marked. So, any dot, no matter the color, means "locked." Here, of course, the problem is ensuring that people both know the rule and agree about which state is the normal state. (Usually, convention says the unlocked and off states are normal, and therefore not marked. But not always.)

Color-coding is not always a simplification. The common color code for "on" and "off" is the colors "green" and "red," respectively. We can overcome the problem of color blindness by making the red be red-orange and the green be blue-green, but it is much more difficult to overcome the problem of large numbers of cases. This is what computer scientists call the "scaling" problem. Something that works well with only a few cases often fails as the number grows.

When there are many different indicators, it becomes difficult to know what to make of the cacophony of red and green lights. The control rooms of nuclear power plants, for example, can have over 4,000 controls and indicators, some of which should normally be on, some of which should normally be off. When looking around such a room, how do the operators know if all the switches are in their correct, normal state?

Signs may be one way of coping with technology: putting up continual signs, reminding people of how things work, instructing and imploring and cajoling people to behave appropriately and to do certain actions, not to do others. We all realize that signs are indicators of poor design. We shouldn't need to post signs. In the ideal world, the design would be so appropriate that proper operation would follow naturally. But in our imperfect world, many people throw up their hands at poor design and try to compensate by using labels.

When Signs Fail

Signs that we put up for our own use can be quite helpful, but the signs of others can be a source of difficulty. It's not easy to keep signs up to date. When they are our own, it doesn't matter much because we know to ignore them, but what about the signs put up by other people? When I walk through an unknown space, how do I know which signs are correct and which are out of date? Figure 3.4 shows two areas with confusing signs, both potentially dangerous.

As shown in figure 3.4, old signs tend to remain in place way past their prime. When I asked why the door of figure 3.4, conspicuously labeled "fire door keep closed," was always open, I was told not to worry, the sign was out of date. Both of the situations shown in figure 3.4 pose safety hazards; the first is that in a fire, the exit sign might lead people to a false hope of escape, and the second is in training people to ignore safety-critical signs. To people who habitually use these buildings and who are aware of the changes, the inappropriate signs may not cause any problems, but to newcomers or infrequent users, they make an otherwise simple thing into something both complex and complicated.

Figure 3.4
When simple things become complicated: out-of-date signs. Was the door on the left once the exit, but when things changed, it was easier to add a negating sign than to get the exit signs removed? The fire door on the right (with the sign enlarged) was encountered on my tour of a major industrial facility. "Not an Exit" photograph taken by Iain Tait (<http://crackunit.com>), who has graciously permitted it to be used here. The "Fire Door" photo was taken by the author.

People, especially busy administrators, often try to rely too much on signs. But signs are like instruction manuals: hardly anyone reads them. My favorite example is a set of signs posted in a civil engineering conference room at Northwestern University (figure 3.5).

Consider two points of view. The first is the administrator of the department, trying hard to maintain sanity in the chaotic academic environment so typical of universities, trying to keep the budget balanced despite pleas from everywhere for more funds. Among the recurring, expensive items is the continual replacement of bulbs for the digital projectors located in all of the department's conference- and classrooms. Those bulbs are expensive. Why do they burn out so often? Professors forget to turn off the projectors after finishing a class, so the projector may stay on for a long time afterward even though nobody is using the system: it could even stay on for days, over the weekend, for example.

Now consider the life of the busy professor. The professor enters the classroom, late for class as usual, and rushes to the podium to set up the slides for the day's lecture. This requires using a key to unlock access to the computer mouse and keyboard as well as using the touch-sensitive display screen to turn on the projector and set it up so that it takes its input from the proper device, either the in-classroom computer or the professor's own laptop.

Note the unusual number of signs: something is clearly wrong here. Some warn the preoccupied professor to read the signs. Most warn the professor to turn off the projector when finished. These signs do little good at the start of the lecture. The professor wants the projector on: the sign is only relevant at the end, so it is ignored for now. But no matter which way the professor looks, new

Figure 3.5
Warning signs don't work. Notice the barrage of signs confronting the person who wishes to use a projector in this conference room. From entrance to exit (read left to right, top to bottom). The fact that there are so many warning signs demanding that the projector be turned off after use, is proof that this method doesn't work.

signs are encountered, reminding the professor to turn off the projector when finished. "Of course, of course," the professor mutters.

At the end of the lecture, oops, the lecture has taken too long, so everyone must run to their next appointment. The professor grabs notes and computer and rushes out the door, only to be confronted with more signs: "Turn off the projector." Too late—the professor is thinking of other things.

The administrator is frustrated. More signs clearly don't work. Announcements in department meetings don't work. Meanwhile, the budget deficits are rising.

This is clearly a situation that requires a design solution, in this case some automation. Many projectors solve the problem this way, turning themselves off if a long period has elapsed with no input signal.

How Experts Can Make Simple Tasks Confusing

Sometimes complexity comes simply from the sheer amount of information that has to be sifted through. All of us face this problem every time we search for information on the Internet. The act of searching is simple, and so is reading a response. But when a simple search yields too many responses to be examined and none of the visible ones appear to be relevant, what is one to do? A simple problem has become complex. It is no wonder that most people simply choose one of the top few items returned by their request and never look beyond that.

The same thing happens with traffic reports broadcast on the radio. I may simply wish to know if there is heavy traffic along one small stretch of highway: a simple question with a simple answer. Unfortunately, in a large city, the announcer has to attempt to cover everyone's need for similar information on their particular roads. As a result, traffic reports on the radio often consist of several minutes of rapid nonstop talking, with the announcer reciting the traffic status of multiple locations, using place names that only the most knowledgeable local citizens are familiar with, describing inbound and outbound traffic—and the experience is so overwhelming that you are apt to miss the description of the traffic you care about.

Recently a friend of mine, Henri Aebischer, sent me an essay about his experiences with weather reports in Britain broadcast by the British Broadcasting Corporation (BBC, called "the Beeb"). I immediately sympathized: same problem.

> The weather forecast on the respectable BBC TV news has become rather extravagant, confusing and, ultimately, quite useless. Unless I concentrate hard on what the presenter says and on where the camera "flies" over the giant slanted map of the UK, I can't figure out what the weather is going to be where I live, in the next twelve hours.
>
> The main problem is that they are trying to cram too much information in a few minutes and through two competing channels. First, there is the voice of the dedicated show person (the Beeb seems to employ an army of them—50 names on their website) whose challenge is to tell us what the weather will be, in the next five days (the prowess of supercomputing), in England (East, Midlands, North East, North West, South, South West, South East), Northern Ireland, Scotland (North and South), Wales

> and Channel Islands, and who tries to avoid repetitions and boredom through flowery language and various gesticulations.
>
> Then there is the supermap of UK, a typical example, in my opinion, of technology misuse. One sees a slanted map of the country, without relief, where the land is sad brown (we have a drought South of London but it's not yet a desert). Some areas are shaded (darker brown); I'm still trying to figure out what that means (cloudy weather?); and rain is shown as blue areas, as if it creates immediate lakes—here's how it looks like. The camera moves over that map and zooms into specific regions as the presenter tells you at the speed of a bullet train what the weather has been (the only sure thing) and what it will be, in that part of UK, today, tomorrow, the day after, day four and day five. Any normal member of the TV audience is overwhelmed and numb after 34 seconds. (Henri Aebischer, email to the author, 2009. Reproduced by permission)

Traffic instructions, weather forecasts: so much information crammed into so little time with hardly a thought as to what it means to the average person. I've listened to many hours of instructions by air traffic controllers to pilots, and to my unskilled ears, they remind me of the weather forecasts and traffic advisories. The differences are that the air traffic controllers speak with a standardized technical language, with special signals to indicate to whom the particular message is directed. Thus, although the controller may issue a long barrage of instructions to several of the airplanes being supervised, each instruction is preceded by the flight number or other identifying name of the recipient. As a result, pilots do not have to listen to everything: they simply "word spot," listening just carefully enough to detect when the name of their flight is called out. Once they hear their flight number, they can switch their attentive state from the secondary task of monitoring to the primary task of full concentration on the following message. To ensure that the message has been received, each pilot is expected to acknowledge receipt and understanding of the instructions. Not only is the information well structured to make it easy to monitor, the pilots are well trained at this task, having spent thousands or even tens of thousands of hours flying.

But what about the rest of us? I shouldn't have to be an expert to understand the weather forecast or the traffic update. The experts who give this information know too much. They have great difficulty understanding the problems faced by beginners.

Even hobbies can be confusing. Recently my wife and I took a "birding" course. We enjoy long hikes in the city, the woods, mountains, and especially along the seashore. Wouldn't it be nice to know what kinds of birds we were seeing? So we signed up for a class on birds. Soon we found ourselves focused on the myriad of details that bird watchers use to distinguish one species from another. Is that a Clark's grebe or a western grebe, or perhaps an eared grebe? (All we cared about was distinguishing a grebe from a duck.) The bird books were not helpful either.

I tried to explain my confusion to the instructor. "These books are all organized by the kind of bird," I said, "which means you have to know what the bird is in order to look it up. Isn't there some guide that organizes them by structure: size, markings, behavior, color, etc.?"

She was sympathetic, but her answer was not helpful. "After a while you learn the birds, so then it's OK," she said. But I need help in the learning! Silhouette, field marks, posture, size, flight pattern, and habitat—those are what the experts claim are the critical features: then why aren't the books organized that way? In this particular case, the complexity comes from the technological limitations of paper books: the rigidity of the bound pages forces the book to have a single, fixed structure.

Fortunately, in today's world of computers and hand-held devices, guidebooks can have whatever organization fits the needs of the moment. Birds can be organized by geography, by color, or by size. They can be organized by their songs and their behavior. Better yet, a person can specify several characteristics and the guide can return all examples that match those characteristics. The guide could help by providing a list of standard identification categories, and as various categories were selected, it could return a list of candidates, or perhaps ask questions to narrow down the possibilities even further. Better yet, these guides already exist: they have changed the way we watch birds.

Note the contradictions. Existing bird guides are easy to explain. They have a fixed, easy-to-understand organizational structure. Look up the name of the bird or species, and there is everything you need to know. Electronic guides are more complex to understand, for some do not even have a fixed organizational structure. These guides are like the Internet: there is no simple organizational structure. To find anything requires search: specify the features that are known and the guide presents a set of possibilities. For the beginner, the electronic guides, even if difficult to explain, are easiest to use. For the expert, the fixed structure is superior.

Reducing Complexity through Forcing Functions

Consider the humble roll of toilet paper (figure 3.6). Even the simplest of things can have a hidden complexity. A single roll, however commonplace, is, well, shall I say, awkward. When the roll empties, then what? When a home or public facility is shared, then social issues come into play. When we remodeled our home, we decided to install a dual toilet-tissue holder so that whenever one roll was finished, the replacement would always be readily available. We purchased a two-roll mount like the one shown in figure 3.7.

We soon discovered that although we now had two rolls instead of one, the problem was not solved. Both rolls ran out at the same time. Sure, it took twice as long before the rolls emptied, but we were still stuck with the same result: an awkward absence of toilet paper. We had discovered that switching to two rolls meant we had to use more sophisticated behavior: we needed a selection rule. Computer scientists call the systematic application of a rule to behavior an "algorithm."

After some self-observation and discussion among a growing circle of tolerant friends, we discovered that three different algorithms for selecting from two visible rolls were in use in the general public: large, small, and random.

Algorithm Large Always take paper from the largest roll.
Algorithm Small Always take paper from the smallest roll.
Algorithm Random Don't think—select a roll randomly.

Figure 3.6
The traditional toilet paper holder. What do you do when it runs out?

Figure 3.7
The dual-roll toilet paper holder. Which roll would you take from, the big one or the small one?

We had assumed that Algorithm Random was the most natural. This would be a poor choice of algorithm. If our selections were truly random, we would choose each roll roughly equally, so that both would empty at the same time. Algorithm Random is not the one to use. To use toilet paper requires some thought.

The most natural algorithm, we soon discovered, was to reach for the larger roll. Alas, consider the impact. Suppose we start with two rolls, A and B, where A is larger than B. With Algorithm Large, paper is taken from A, the larger of the two rolls, until its size becomes noticeably smaller than the other roll, B. Then, paper is taken from B until it gets smaller than A, at which point A is preferred. In other words, the two rolls diminish at roughly the same rate, which means that when A runs out of paper, B will follow soon thereafter, again stranding the user with two empty rolls. Algorithm Large is what a computer scientist would call a "balanced usage" algorithm. This is not what we want with toilet paper usage.

What about other people? I took this question on the road, asking audiences which roll of the two shown in figure 3.7 they would use—the big one? The small? The majority invariably revealed that they would choose the large one.

Algorithm Small is the proper choice. With Algorithm Small, paper is always taken from the small roll, so it gets smaller and smaller until it runs out. Then paper is taken from the other one, which is full size at the time of the switch.

Yikes. We never realized that thinking was required to select the roll. The difficulty here, though, is that the natural inclination to select from the larger roll defeats the design goal.

The dual-roll toilet paper holder of figure 3.7 illustrates a case where perfect visibility is harmful rather than helpful. When the size discrepancy between the rolls is visible, it leads many people to precisely the wrong behavior. In design, the same principle that works in one situation may be precisely wrong for another. Here is a case where a seemingly simple device has hidden complexity.

There is a design solution to this problem. Instead of a dual-roll holder where both rolls are equally available, the holder should enforce a serial constraint: the second roll should not be available until the first is depleted. This is what I called a "forcing function" in *The Design of Everyday Things*. And indeed, many commercial toilet paper holders exist that apply appropriate forcing functions. Quite often the new designs solved one problem at the expense of others, which gave rise to yet new and different solutions.

In some commercial holders, when the bottom roll is depleted, pushing down on the empty holder will release the upper roll. However, the upper roll, if any, is quite invisible behind the solid metal container. Is there a spare roll? There is no easy way to tell. Do the makers of such holders expect the janitorial staff to open up the containers of all of the toilets they must service every day to determine if there is really a spare roll waiting? I doubt it. So this design has, in turn, set off a rich panoply of alternative designs. Some have transparent sides so that the spare roll is visible, even if not accessible, others have the spare roll visible, but blocked from access by the roll in use.

The toilet paper holders illustrate several points, including the need to communicate (by making visible the presence or absence of backup rolls), and the power of design that constrains behavior to be appropriate, (always taking from the smaller roll until it is completely empty). This is a forcing function: the correct behavior is the only possibility. They also indicate that even the simplest of designs may have social implications. Even toilet paper holders are social objects.

Human behavior can be deceptively complex: social behavior is even more so. We must design for the way people behave, not for how we would wish them to behave. People function well when the devices they are using make things visible, provide gentle nudges, signifiers, forcing functions, and feedback.

The solution to the toilet paper problem was an appropriate forcing function, a constraint that naturally caused the correct behavior. One nice property of a properly designed forcing function is that the correct behavior minimizes the need for problem solving or decision making: the person is gently nudged toward the appropriate behavior.

Complexity is a fact of life, so we must learn to deal with it. Sometimes the devices we must use are complex. Sometimes the devices are simple, but circumstances make them complex. We need coping behaviors that help us manage the complexities of life.

Put the knowledge in the world. One simple solution is to add hints and suggestions. If airlines can put lines on the ground to help their crews, we can do the same for ourselves. Use whatever tools work best: sticky dots, forcing functions, posted instructions, or even replacing simple technologies (e.g., books) with more powerful, organizational technologies (e.g., electronic bird guides).

Use the knowledge in the world. The same is true for behaving: in a strange place, how do you know how to act? Look around; follow the examples of others: do what they do. In a strange culture with unknown language, how do you order a meal? Look around to see what others are eating, and order whatever looks interesting: all you have to do is point. Use the knowledge that others have placed in the world.

Life can be complex, but we can learn to cope. Sometimes it is the technologies that provide the complexities; sometimes even simple technologies yield complexity when they come in too many sizes, shapes, and forms. Sometimes it is technology that will rescue us from complexity, whether through automation, better design, or dynamic organizational structure that can configure itself to provide the information we need just when we need it.

Figure 4.1

Crowds as a social signifier. Did the train already leave? The state of the train platform provides the answer. Here, the presence or absence of waiting passengers serves as a social signifier, signifying a train that has not yet arrived or one that has already left. Social signifiers are not guarantees, but they are strongly suggestive.

4
Social Signifiers

Despite the complexity of the world, most of us manage to function quite successfully. We even manage well in novel situations, where we do not have the benefit of prior knowledge or experience. In part, this is through the subtle cues provided by the activities of others. People's actions have side effects, leaving behind traces and trails of their activities that enable us to retrace their steps. Most of this is done without conscious awareness, but the side effects are important social signals: what I call "social signifiers." Social signifiers allow us to navigate in otherwise complex, potentially confusing environments.

A "signifier" is some sort of indicator, some signal in the physical or social world that can be interpreted meaningfully. Deliberate signifiers are intentionally created and deployed. Incidental signifiers are accidental by-products of activities and events in the world. Social signifiers result from the behavior of others. Designers use deliberately placed signifiers to aid appropriate usage.

In the vocabulary of design, signifiers are often called affordances or, more precisely, "perceived affordances." This is actually my fault, due to my introduction of the term in *The Design of*

Everyday Things. Unfortunately, the term affordance has deeper meaning than that of a signal: an affordance does not even have to be perceivable. The introduction of "signifier" is intended to make the design vocabulary more precise. (I discuss this more fully in chapter 8.)

Suppose you are rushing to catch a train. You hurry to the train platform at the station: did you miss the train, or has it not yet arrived? How can you tell? The state of the platform serves as a signifier (see figure 4.1).

In figure 4.1, the two signifiers are not symmetrical: the crowd of waiting people can be considered strong evidence that the train has not yet arrived, while the empty platform suggests that the train might have already left. Of course, if the platform is empty, it is possible that the train has not yet arrived, but nobody else wanted to catch this particular train. In a busy train station, say, one at the center of a large city, there are many trains arriving at frequent intervals, so the presence of a crowded platform provides no information about a particular train. Instead, it signifies the usual rush of commuters. Even here, however, the presence or absence of a crowd that differs from expectations can be significant: "Why is it so crowded?" you might wonder, surprised at seeing a crowd on a weekend in the middle of the day. "What is happening?" Similarly, the absence of people in what should be a busy period is significant, even if exactly what is being signified remains elusive. The exact interpretation of a signifier depends on other knowledge, such as the time of day (rush hour versus a quieter time of day).

Signifiers indicate critical evidence, even if the signifier itself is an incidental by-product of the world. Many signifiers are deliberate: deliberately designed and installed to be informative. Some are unintentional side effects. A shadow is a side effect of the presence of a person or object in front of a light source, but when we see the shadow we immediately infer the presence of the object. A shadow is both incidental and natural: it was not designed or put into place, but rather is a natural result of the physical world.

Does this restaurant want you to seat yourself or wait for a host? Look around. How to eat an unfamiliar food, or use unfamiliar utensils? Watch others. What path to take through the snow? Follow the footsteps. Have trouble navigating a dense crowd of people? Follow closely behind someone else. What book to read, movie to see, restaurant to select? Ask what others have done, especially friends whose tastes you believe similar to your own. Crowds, society, other people have a lot of wisdom, which they share sometimes explicitly, as when they answer questions or post messages, sometimes implicitly, as when the impact of their activities creates signals that can be interpreted: footsteps in the snow, crowds at a restaurant, even just wear and tear.

The tendency to follow other people's actions gives rise to a common children's trick: stand on the street corner, look up and point. Soon others will join you, staring upward at nothing. It doesn't take long before there is a crowd. Why? Because the actions of others are usually informative, alerting us to important or interesting aspects of the world. Why shouldn't we look where others are looking? We might find it interesting. The children's trick exploits this natural behavior.

In the physical world, social signifiers are themselves physical, but not in the virtual world of electronic interaction and communities. Nonetheless, the traces of activity in the virtual world are just as powerful as those in the physical, as the large number of "recommender" systems on Web sites, social networks, and location- and theme-based message systems testify. Recommender systems capitalize on the trails people leave behind by their activities: "people who have liked this item," they will inform you, "also liked these other items." This enables you to follow their trail if you consider yourself similar in your likes and dislikes. The browsing, reading, and purchasing behavior over electronic media is simply the virtual equivalent of physical trails, such as footsteps in the snow.

Biologists and those in the field called "artificial life" call this phenomenon "stigmergy": indirect coordination, usually through traces of earlier activity. Animals leave footprints; ants leave chemical trails. Complex animal structures such as termite mounds, wasp nests, anthills, beaver dams, and the hexagonal honeycombs of bees get built without the need for explicit, intentional goals to do so. Instead, the trace of the previous activity constrains and directs the future activity. The end result is the construction of complex structures and behavior through a self-organizing process, without the need for a goal or leader. Good concept, but what a strange word for it: stigmergy.

These traces are signifiers, mostly unintentional and accidental, although it is possible to make a case that evolutionary forces "deliberately" produced the chemical trails left by ants and other animals and even evolved animals so as to be able to use the traces of previous activity to guide in their construction of nests and hives. After all, each animal is genetically endowed to construct their special, unique structures.

In human behavior, the unintentional signifiers can be exploited by designers. One group of researchers noted that:

> bindings of cheap paperbacks bend and crack in a manner that allows one to find the last page read. In an auto parts store, the most often consulted pages among many linear feet of catalog are identifiable by smudges, familiar tears, and loose pages. The smudges, tears, and loose pages index to information users are likely to consult. Switching from auto parts catalogs to door handles, the polished part of an otherwise patinaed brass door handle shows where others succeeded in grasping it. The best recipe cards in a stack are often dogged-eared and stained. (Hill et al. 1992, p. 6)

From observations like this, they designed a computer system that deliberately left marks wherever people stopped to read or edit: read marks and edit marks, they called them. These marks mimic the smudges left behind on a physical book by readers and editors.

Cultural Complexity

Consider the salt and pepper shakers of figure 4.2: which one contains the salt? I have asked this question of audiences around the world and the results are invariably the same: half the audience thinks the salt is the one on the left, half thinks it's on the right. When I ask for their reasons, both sides have strong reasons for their belief, the most common being the number or size of the

Figure 4.2
Which is the salt? Each shaker is simple, but to know which contains salt requires a combination of practical and cultural knowledge. Moreover, both the person who fills the shaker and the person who uses it must be in agreement. This is why many of us test first, shaking some of the contents onto our hand or plate before applying it to the food.

holes: "salt is on the left because it has more holes"; "pepper is on the left because it has more holes."

Which is correct? It doesn't matter. What matters is what the person who filled them believed.

Salt and pepper shakers illustrate yet another source of complexity: culture. They may seem to be simple devices, but they are part of a social system. One person fills the shakers, someone else uses them. A good designer thinks about these things and provides clues—signifiers—as to appropriate use. This requires a special talent: empathy. Designers must place themselves in the position of those who use their designs, and then provide the information required for proper usage, but without destroying the aesthetic beauty or the functions, or increasing the cost. That's the design challenge: managing the tensions. Less empathetic designers (and sometimes I think these are the majority) focus only on one or two aspects of a design, perhaps appearance, perhaps engineering practicality, perhaps cost. The result is products like the salt shaker, the modern television set, or the Roland piano described in chapter 1. For the designer to say "but everyone knows which one is the salt" is inexcusable. To use these salt and pepper shakers properly requires that everyone have the same knowledge. It requires social synchronization, and social synchronization is the most difficult of all activities.

I have questioned waiters and restaurant managers about their methods. I am always given authoritative answers about which is salt, which is pepper, except that different restaurants have quite different answers. Once, in a fancy restaurant in Amsterdam, I was also told that when the salt and pepper shakers are put on the table as a pair, the salt shaker is always closest to the entry door to the room. I queried random waiters about this and they all agreed: obviously they were well trained. Then I walked around the several dining rooms of the restaurant, testing this new knowledge. I found a few tables where the rule did not seem to work. I quickly found the manager and asked him what this meant. "Oh," he said, quickly readjusting the positions of the shakers, "these are wrong." Rules are useful, except when they are not followed.

There are several morals to this story, all of which illustrate the reasons we have so much difficulty with our technology. First, to understand how we function in the world we must understand how we interact with one another.

Second, in the real world, where so much is unknown or uncertain, where so much is not under our control, the best tactic is usually "proceed with caution." If possible, do an experiment. Thus, most people, when confronted with the opaque metal containers of the salt shakers in figure 4.2, sprinkle some on their hands to see which comes out, salt or pepper.

Third, good design can make the entire problem disappear. Thus, the shakers shown in figures 4.3 make their contents visible, so no question need be asked.

The shakers of figure 4.2 lack social graces. They require that both the people using and filling them share the same cultural knowledge, and because most people don't trust that this is true, they must engage in problem solving simply in order to put salt or pepper on their food, either by experimentation—trial and error—or by watching the people around them to see which shaker they use.

There are many ways to add sociability to these devices, and there is a wide variety of salt and pepper shakers in the world that

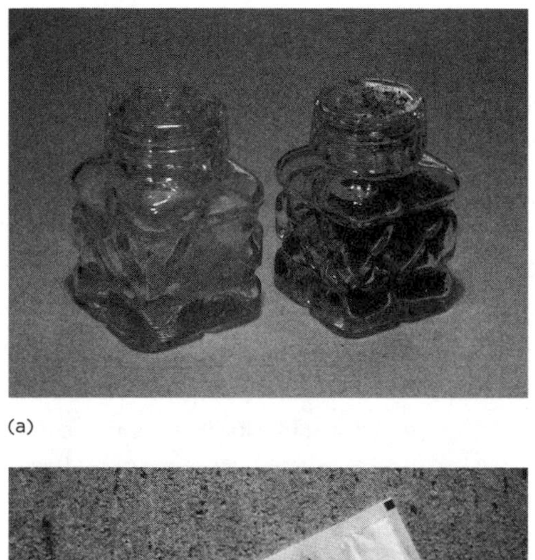

(a)

(b)

Figure 4.3
In (a) we see a design that makes it easy to tell which is the salt. You don't have to know anything: you simply look. (b) shows salt and pepper shakers used by United Airlines, demonstrating that even inexpensive containers can be both attractive and understandable.

have done so. Some popular ways of overcoming the confusions illustrated in figure 4.2 are making the shakers transparent so the contents can be seen (as in figure 4.3a), labeling the shakers, as in figure 4.3b, and by arranging the holes into an "S" or "P" shaped configuration. These alternative designs put the knowledge in the world, making operation self-explanatory. Now, the presence or lack of sociability in these simple items is not a major component of life. But the examples serve to illustrate the principles, showing that they apply to almost everything we interact with. As items become more complex, with sophisticated mechanical, electronic, and communication technology, the same principles apply.

Although it is true that the cultural complexity of judging which container has the salt is readily overcome by appropriate design of the shaker, the point still remains: cultural explanations are largely hidden from sight. This leads to needless complexity and potential confusion, error, and embarrassment unless the designer has gone to special efforts to provide signals for appropriate behavior.

Functional design—that is, the part of design that makes the objects around us usable and understandable—is primarily about communication. Fail to communicate properly and you get frustration at best, accidents and disasters at worst. Proper design can minimize the need for arcane knowledge or experimentation; but we live in a society of people, and so to live comfortably in the modern world, we must come to understand the role played by social interaction, by groups, and by culture.

Social Signifiers: How the World Tells Us What to Do

Social signifiers, such as the presence or absence of people on a train platform or painted lines on the street, are all examples of signaling systems. The role of signals has long been of interest to biologists, anthropologists, and other social scientists interested in how animals and people indicate information of interest to one another. The academic field known as "semiotics" is devoted to "the life of signs within society." In some cases, the signals are provided by evolution, as when an animal indicates its strength by its size, its roar, or perhaps its horns or antlers. Sometimes it is indicated by behavior, such as many courting rituals of animals or even the "handicapping" behavior of strong animals, bragging about their ability to overcome apparent handicaps. The long tail of the peacock is thought to be one such signal, for it would appear to hamper flight, but one current theory holds that the tail is really an act of boasting. The long-tailed male is claiming that it is strong enough to overcome that handicap. Another example is the gazelle that does not flee when it spots a lion, but rather jumps up and down in the air, as if to say "ha ha, you can't catch me." It appears that the wise lion believes it and instead chases after the weak or the young, who signal their vulnerability by being the first to flee.

The signifier is an important communication device to the recipient, whether or not communication was intended. This makes my use different from that of many theorists who want communication to require intention on both sides, that of the sender and the receiver. For the purpose of surviving in the world, it doesn't matter to an individual whether the useful signal was deliberate

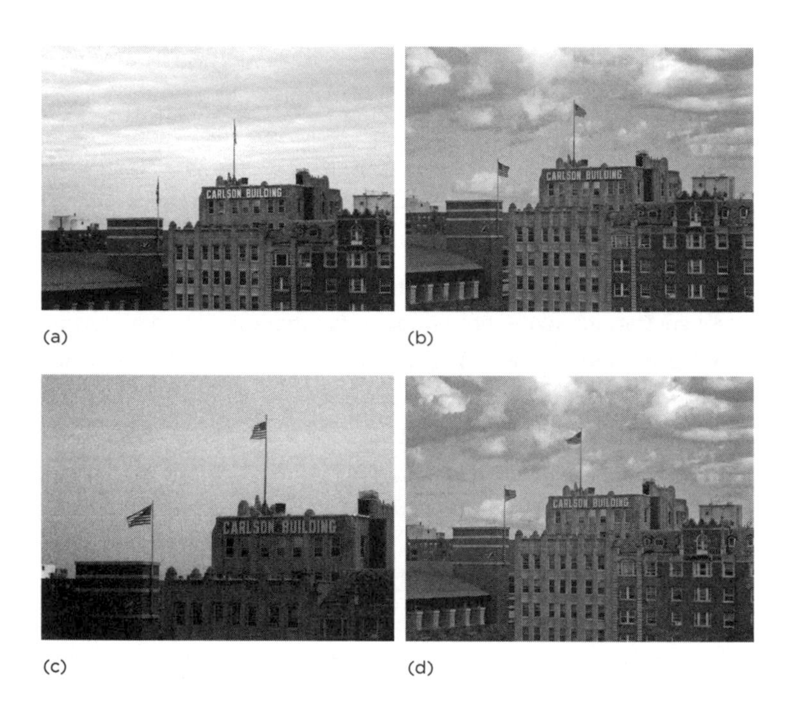

Figure 4.4
Flags as natural signifiers. When I look out of the window of my home near Northwestern University, I look at these flags to judge the weather. If I see the scene in (a), I know it is a calm peaceful day. If I see (b), there is steady wind from the north whereas (c) shows a steady wind from the south. But what do I make of (d)? The flags are just across the street from one another, yet they are blowing in opposite directions. Maybe I should stay home—the weather will be gusty and unpredictable. (The photographs are all genuine: the flags actually do this.)

or incidental: to the recipient, no distinction is necessary. Why should it matter to me whether the flag was placed as a deliberate clue to wind direction (as is done at airports or on the masts of sailboats) or whether it was there as an advertisement or a symbol of patriotism (as in figure 4.4)?

Flags can serve as natural signifiers, but because they convey useful information, they are also deployed as windsocks at airports and other locations to indicate wind direction and speed. Thus, figure 4.5 shows an artificially constructed flag, a windsock, used as a deliberate signifier.

Whatever their nature, deliberate or accidental, signifiers provide valuable clues as to the nature of the world and of social activities. For us to function in this social, technological world, we need to develop internal models of what things mean, of how they operate. We seek all the cues we can find to help us in this enterprise, and in this way, we all act as detectives, searching for whatever guidance we might find. Sometimes thoughtful designers provide the clues for us. At other times we must use our own creativity and imagination.

Social Signifiers in the World

Figure 4.6 shows the amazing variety of deliberate signifiers intended to ensure our safe traversal of city streets. The multiple varieties of painted lines are cultural signals, for their meaning can be guessed from the context, but they are unique to British traffic culture. Americans do not understand the meaning of the zigzag lines. Even the colored signals of traffic lights are culturally

Figure 4.5
A windsock as deliberate, artificial signifier. When windsocks are deployed at an airport there are rigid specifications for their design so that pilots can tell wind speed by how fully the sock is extended. The photograph shows a windsock only slightly extended, which means the wind is less than 15 knots (17 mph).

Figure 4.6
Street crossing in London. Note the variety of signifiers: a fence to constrain pedestrian traffic, street dividers to constrain automobiles, signs to remind pedestrians to "look right" before crossing, and a variety of painted lines—light dashes, heavy dashes, wiggles (zebra marks), arrows, and solid lines. Also a traffic signal that flashes green or red. Notice the pedestrian, apparently oblivious to the marks, walking in complete violation of their intended meanings.

determined, but in this case, fairly universal in using red for stop and green for go, with minor variations in the coloring to accommodate the belated recognition that roughly 10 percent of males cannot distinguish red from green (a fact well known to sensory psychologists and ergonomicists, but not to the early inventors of traffic signals). The physical barrier in the photograph, the fence, is more likely to constrain behavior than the arbitrary cultural signs of lights and painted lines, as can be seen by the pedestrian crossing the street in violation of the painted lines. Note too that even the proper location of the vehicles is arbitrary and culturally assigned: the truck is driving on the left side of the road, proper for London, but it would be a violation of traffic codes in most of Europe.

Social signifiers cannot compel appropriate behavior. The pedestrian in figure 4.6 is crossing the street in direct violation of the numerous social signifiers, including a traffic light and the lines intended to mark the proper location for crossing. Social signifiers are conventions, sometimes suggestive and helpful but completely voluntary, and at other times legally defined and enforced through police and the legal system. But many societies are tolerant of minor violations of these signifiers, so the aberrant pedestrian is seldom penalized whereas a similarly aberrant automobile would be targeted. Why? Social signifiers depend on social interpretations, social systems, and cultural structures.

Social signifiers, moreover, are highly responsive to levels of authority. Imagine that you are a lowly ranked guest at a formal dinner, one of those stiff affairs where the table is set with an apparently endless array of knives, forks, and spoons. Pick up the wrong utensil and prepare to be embarrassed. Pick up fried chicken with your fingers, a behavior that is perfectly normal at casual dinners and picnics, and be prepared to be banished from the table. Obviously, the proper way to behave in these circumstances is to watch the people around you and do whatever they do.

If, however, you are the host of the dinner, or perhaps a visiting dignitary, then you may do whatever you please. Eat the salad with your fingers? Don't be surprised if some of the other dinner guests start doing the same: they too believe in following the leader. A quick search of the Internet with the string "drinking from finger bowl" indicates that this can indeed be a major problem, for many a guest, never before having seen a bowl of water next to the dinner plate, assumes it is for drinking, not for dipping one's fingers. Once, so goes a common British legend, when a guest of Queen Victoria of England made this error, the queen followed suit so as not to embarrass the guest.

The Queen Victoria story is probably false, likely yet another example of how cultural knowledge spreads as urban legends, based more on the allure of the tale than its accuracy. Nonetheless, the cultural point is true. People do unwittingly drink from finger bowls and some advisors do suggest that the host follow the example to avoid embarrassment. This is an interesting example of deliberately producing apparent appropriateness: when one person violates standard cultural rules of behavior, others might follow the transgressor to make the act appear to be normal and appropriate.

Painted traffic lanes and pedestrian crosswalks are both deliberate and explicit social signifiers. Once one is attuned to them, they are ubiquitous: there is a large number of workers employed to paint and otherwise place these deliberate signifiers of proper behavior, although I suspect the number of people who understand their meaning is less than those who design and place them in the world imagine. I have found these signifiers inside airports, on the runways and ramps of airports, in hotels, in hospitals, and almost anywhere that people must be guided to stay within proper lanes, to stop at proper locations, or even to swim, jog, or bike in appropriate places. Deliberate signifiers are the easiest to find, for they are usually designed to be visible, the better to guide behavior.

Cultural complexity often arises when the cultural norms are vague and ill learned. Do you know what to do with your napkin if you must leave the table at a fancy dinner? It turns out that there are strict rules of proper napkin-leaving-behind behavior. I encountered this lovely tidbit on the *O: The Oprah Magazine* Web site:

> Place the napkin in your lap shortly after you sit down. If you excuse yourself during the meal, leave your napkin—folded or unfolded—on your chair and push your chair in. When you've finished your meal, fold your napkin and put it to the left of your plate, a signal to the waitstaff that your last dishes should be cleared.

This is a wonderful example of a deliberate, explicit social signifier, both in the explicit signal it sends, but also in the very social nature of the signal. How many dinner guests know this ritual? How many of the waitstaff? I, for one, did not know about this rule until I found it on the Web site in preparation for writing this chapter. Social signifiers, even the most deliberate, even the most explicit, only work if the relevant others know about them.

If intentional and clearly defined social signifiers are problematic, what about ones that are deliberate, but not at all visible? Consider waiting in line. Here the social signifiers are not explicit. In many cultures, the presence of a line indicates an order in which people are to be served, and the physical presence of the line acts as a deliberate signifier of queuing behavior. In particular, it is inappropriate to enter the line at any location but the end, and the violator is promptly corrected. Some cultures allow invitation of friends into early positions, others do not. And some cultures do not follow the neat, orderly line behavior.

I witnessed the clash of cultures around this implicit social signifier at an amusement park in France where the patrons represented a mix of many European cultures. Some cultures waited patiently in an orderly fashion while other cultures assumed it was proper to move as quickly as physically possible to the front of the line. It took constant interactive intervention by park employees to prevent fights during these culture clashes. (I return to this in my treatment of waiting in chapter 7.)

Sometimes apparently obvious signifiers are not as they appear but are caused by completely different events. As a result, the obvious, immediate interpretation can be wrong: call these "misleading signifiers." A good example of a misleading signifier is the slowing and even stopping of traffic on a crowded highway. To most people, the slowdown is an indicator of problems ahead; perhaps an accident has blocked the traffic lanes. But sometimes a traffic delay is a misleading signifier, and is actually an indicator of a nontraffic event.

How can a nontraffic event cause traffic to come to a halt? Suppose a house catches fire, with flames visible to the automobiles on an adjacent, crowded highway. Drivers slow down for a brief glimpse of the spectacle, which means that the cars behind the slowing vehicles must also slow down to avoid collision. Each succeeding car slows in turn, but with increasing delay because of the time it takes for each new driver to notice the slowdown. The result, well known to traffic engineers, is a traveling wave of slowness, propagating backward from the site of the fire. The slowdown is more pronounced and longer lasting the farther back it propagates. Finally, at surprisingly distant locations, kilometers or miles away from the fire, all traffic halts. "Must have been an accident," the drivers are bound to think, taking the stoppage of traffic as a signifier of serious trouble ahead. This is a misleading signifier, in this case a particularly interesting bit of emergent behavior, much beloved by traffic engineers and professors.

How many of the events that we witness and interpret as meaningful are in fact accidental signifiers of something completely different? The stopping of traffic due to a minor slowdown far ahead is both an accidental signifier and also an example of emergent behavior that results from the collective activities of many even though no plan was in effect.

Social signifiers are valuable clues to the working of the world even though they are sometimes ambiguous, sometimes misleading. But they provide a powerful tool in our arsenal of clues that help us make sense out of an otherwise complex world, helping us learn how to behave by observing the behavior of others, both directly and through their unintended but informative side effects.

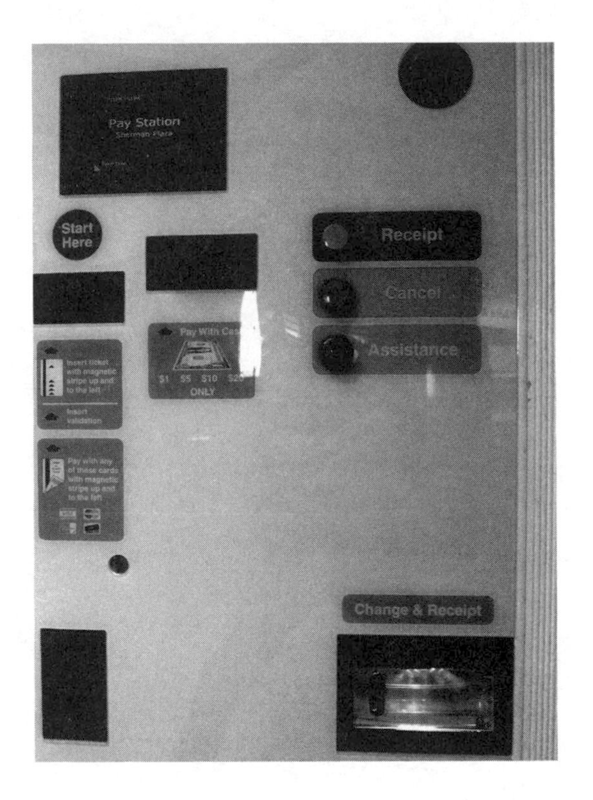

Figure 5.1
"Stupid machine." Actually, it is nothing special. Just a normal parking garage pay station that normally works well, speaking in a polite human (female) voice to its patrons. But when things go wrong, it can't cope, and its normal, pleasant self is replaced by inhospitable, unsociable buzzing sounds.

5
Design in Support of People

"Stupid machine," I heard a woman shouting as I walked through the lobby of the building. She had parked her car in the garage and had now returned, ready to drive away. But first, she had to pay for the parking, using a machine thoughtfully set up in the lobby next to the elevators that would take her up to her car. All she had to do was insert her parking ticket into the machine, which would compute how much money she had to pay. Then she could pay by cash or credit, and the machine would validate the parking pass and return it to her. She would now be able to drive out within fifteen minutes without having to pay extra.

The woman inserted her parking ticket and paid, but then never received the validated receipt, which she needed to drive out. There was nothing to guide her at this point. From her point of view, she had done everything right, but the pass wasn't disbursed. What could she do about it? "Stupid, stupid," she muttered, kicking the machine. She pushed a button: "bzzz" answered the machine. "It won't give me my ticket," she yelled to nobody in particular, pushing more buttons and getting more buzzing sounds in response. You can see a photograph of the machine in figure 5.1.

Machines certainly do act stupidly, but then again, they are simply machines. Why would we expect otherwise? Even "intelligent" machines are not very smart by human standards. Machines have no way of understanding the actual situation, or the context. They can only deal with events their designers have considered, which means they cannot deal with the unexpected. But unexpected events should be expected to occur. The problem is that when they do occur, they are always unexpected. The designers of the pay station were aware of this problem, which is why they had placed a large button labeled "Assistance" on the right side of the machine. At the pay station shown in figure 5.1, pushing the button causes a human response over the loudspeaker, and when necessary, the attendant comes down from the third-floor office to the ground floor to put human ingenuity to the rescue.

Machines can often simplify our lives by taking over some of the routine, mundane parts of everyday life. This parking pay station is a good example. When it works, it makes it easy to reclaim a car from the very large (twelve-story) parking structure at any time of day or night. The problems arise when the machine doesn't work: then the machine adds complication.

Our lives are made more complicated because of machines that cannot cope with the real complexities of the world. Design of machines that can deal with all the complexities, and especially with the unexpected, is not going to be possible for a long time. Still, there is a lot that can be done to help. One sensible approach is to do precisely what the parking machine does: make it very easy to ask a human for assistance. A second problem has to do with the design philosophy often followed by the designers of these machines: the designers lack empathy with the people who must use them. These are problems that can be overcome.

We need designs that can deal with the unexpected in a helpful manner. It's perfectly reasonable for a machine to fail. After all, people make mistakes frequently. But when there are mistakes and failures, the sociable human deals with them: apologizes, tries to fix things, and is helpful and courteous throughout. That's the general attitude our machines should reflect.

This problem of unresponsiveness happens frequently when dealing with bureaucracies. Your request might be blocked or delayed by hidden, unknowable backroom activity. Even if the people you are conversing with are friendly and cooperative, they may be unable to offer any help except to apologize. This is an unsociable interaction. You get no explanation, no understanding of what the issues are, no understanding of why you cannot be helped. Sometimes the employees are just as frustrated as the customers. I can imagine that when clerks get together, they complain about the rules and regulations that prevent them from helping more effectively. The problem here is not that either the clerk or the customer is behaving unreasonably, it is the lack of sufficient information: the employee doesn't know what is going on behind the scenes of the bureaucracy. Everyone ends up frustrated.

When we deal with machines, there is a similar lack of engagement. The problem is that the designer usually focuses on correct behavior, when everything is working properly and the customer acts just as intended. When this happens, everything runs smoothly: the machine works and the customer is satisfied. What happens when things go wrong? Often, the machine cannot sense that there is a problem, so it keeps requesting the next operation, unaware that the customer is stymied, unable to comply. This is what happened with the irate woman at the parking pay station.

The problem is especially acute when the design is done by engineers who view the world from their logical, sensible perspective. From their point of view, people get in the way. "If only we didn't have all these people around," I have heard engineers tell one another, "our machines would work just fine." This mindset is often present in the people who put together the machines: programmers, engineers, and system administrators. When they are forced to accommodate the actual behavior of people, they talk about making their designs "foolproof" or "idiot proof."

Ever try to make a telephone call and wonder if anything was happening? Silence. No clues. So you hang up and try again. When people complained that they could not tell if the system was working without some sounds, the engineers got annoyed. "We can't win," they exclaimed. "People complained about noisy telephone circuits, so we went to great effort to make them completely silent, and then they complained about that." So the engineers made the lines perfectly silent and then added noise back in. But they showed their disdain by calling it "comfort noise." That's an insult. I call it "meaningful feedback." It's not comfort, it's essential.

Ever hear of a "confidence monitor"? Whenever I give talks to large audiences, I stand on the stage facing the audience, usually with bright lights blinding me so that I can't see anyone. When I show pictures, I can't see them either because they are projected somewhere behind me. Speakers have complained that they needed to be able to see what is being shown. Sometimes inexperienced speakers will turn their backs to their audience in order to see what is on the screen and spend the entire time talking to the screen.

There is a simple solution: place a monitor in front of the speaker, so the speaker can face the audience, glancing at the monitor when necessary to check that the image on the screen is the intended one. This solution is accepted practice for large, professionally run presentations. Sometimes the monitors are on the floor of the auditorium, either on the floor of the stage or just in front of the first row of seats. Sometimes the same purpose is served by having a projector display the slides on a big screen at the back of the auditorium. These provide valuable feedback, allowing the speaker to see what the audience sees.

An unsolved problem is that speakers lack confidence in the machines. All too often we have seen things fail, pictures fail to display properly, and videos that don't operate. As a speaker I don't for one minute believe that all my photographs are actually going to show up on the screen. I've given up trying to show videos: they only work during practice. During the real talk, they are apt to sputter and crash. Yeah, I need confidence: I need confidence that the machines will work properly. Call it reassurance. Call it trust. But don't call it comfort noise or confidence monitors, or idiot proofing or foolproofing.

I've collected numerous examples of misunderstanding, miscommunication, and mangled interactions between people and bureaucracies, people and machines, and people and other people, all similar in spirit to the situations discussed in this chapter. The problem is the lack of sociable rules on the part of the designer. These experiences have led me to believe that we need a new mindset for designers: sociable design. I once described the need this way:

> Engineers and programmers, even intelligent, well-meaning ones, have grown up taking the machine's point of view. But these people are experts at the inner workings of technology. These are not the ordinary people for whom we design our systems. Nonetheless, because of their skills, they dominate the technological community. These are the people we need to convince.
>
> I recommend changing the battleground. Bring it back to human terms: Ask for compliance and tolerance. Those are new concepts for designers, but as concepts go, they're easy to understand. Ask our engineers, programmers, and fellow designers to aim for compliant systems, tolerant systems.
>
> As it stands, we must accommodate technology. It is time to transform the technology to accommodate us. (Norman 2009a)

I first got to try these ideas when I served as a consultant for a company developing a radically new version of its Internet software application for American income taxes. Talk about complex tasks: the rules and requirements and the number of different forms required are difficult even for experts. Our program tried to help by providing confidence and reassurance. People could enter the required information in any order they wished. They could skip steps if they weren't ready. And at every step, there was visual confirmation of the action taken, the tentative result, and always, a "more" button that could be pressed to explain why that particular step was required and what to do if there was any problem.

The tax program made it to market only briefly: it was stopped for internal reasons related to the company's business organization. But while it was available, it received very high reviews for its comforting, assistive approach. That was my first attempt at sociable design (at the time, my client called it "emotional design").

The lesson to be learned here is that it is time to socialize our interactions with technology. What is needed? Sociable machines. Basic lessons in communication skills. Rules of machine etiquette. Machines need to show consideration for the people with whom they interact, understand their point of view, and above all, communicate so that everyone understands what is happening.

Reticulating Splines

Reticulating splines
—Screen display from a computer backup program

I was in my home office backing up my files, using one technology to protect me against the potential failures of another technology, hoping to protect my priceless computer files against computer failure, or for that matter, fire or earthquake. My backup program (its name is Mozy) talks to me, giving continual progress reports, letting me know that it is working hard on my behalf, doing wonderful, mysterious, complex, and no doubt absolutely essential things to protect me from unknown disasters that might befall my precious manuscript.

First it tells me that it is scanning my files, then connecting with the server located in some remote, but presumably safe, location. It keeps me in touch with everything it does. At some point it tells me it is "*reticulating splines.*" The very perplexity of the technical jargon is strangely reassuring, suggesting that I don't have to bother to think about such complex things but, instead, can leave it to the experts, which in this case is a computer program on my home machine that is talking to a mysterious "server" located somewhere in the mysterious cloud of servers scattered across the world. All I need to know is that it is working hard to store all my data in remote locations, so that even if my house burns down, even if California slides into the ocean in a massive earthquake, my data are safe.

But what does "reticulating splines" mean? The manual for the backup program did not say. I went to the Internet and searched: I got around 40,000 responses. The phrase "reticulating splines" turns out to be an insider's joke. The game developer, Will Wright, said that he had inserted the phrase into his computer game SimCity 2000 because "it sounded cool." The phrase has since persisted, showing up in games ever since. And, of course, in my Mozy backup program. But wouldn't you know it, we have reached the stage of technological social interaction where there is normally so little communication, so few social skills, that even a nonsensical phrase is reassuring. "Don't bother your pretty little head about this," my technology condescends to tell me, "I'm on the case, reticulating those pesky splines."

We are dependent on machines and systems that we no longer can comprehend. Whether it is the international banking scene, the management of trade, the scheduling of freight and passengers, or even the fare systems of airlines, the rules are so complex that no single person can hope to master them. The operating system of a home computer may contain over 50 million separate lines of commands.

The lack of comprehension between us and our technology isn't just one-way; it goes several ways. The technology does not understand us, nor does it even try. When things go wrong, the lack of information makes it impossible to know what is happening. Our technological world is increasingly unsociable. I have seen world leaders in technology stymied by relatively simple problems because of the lack of information.

Machines, of course, don't have intelligence, although the engineering community is working hard to try to give them some. But with or without intelligence, they need social manners, which is something that is seldom considered. The burden is actually on the designers, not the machines, because the intelligence, courtesy, empathy, and understanding are put in there by designers and engineers. Of course, to us everyday people who must interact with the machines, we see the machine, not the people who designed them. To us, it is the machine that lacks understanding, that causes our frustration, and that is at fault.

The Mismatch between Goals and Technologies

People usually do things for some higher-level goal. The individual tasks that make up an activity are steps toward that goal. For example, we might have the higher-order goal of an enjoyable evening with friends, which has within it the activity of cooking dinner for them, itself a high-order activity. Cooking as an activity is made up of many lower-level activities that may involve yet lower-level tasks. Sharpening a knife is a subgoal of chopping the vegetables, which in turn is a subgoal of preparing some dish; but these activities are all much less important than the high-level goal of a pleasurable social evening.

Although we often perform activities in pursuit of some higher-order goal, the tools that we use are usually specialized. We don't have tools for the important goals, only for the smaller components.

The separation between tools designed for specific tasks and our goal of high-level activities does not lead to difficulties with mechanical tools. The reason is that these tools have stable, understandable behavior. People are able to predict the behavior of the tools and compensate for discrepancies between their high-level needs and the capabilities of the tools, ensuring that each low-level task fits neatly into the high-level goal.

With intelligent tools, problems arise because there is often a mismatch between the behavior and expectations programmed into the machines and the expectations and behavior of people. We often need to change our goals part of the way through an activity. Or we may decide to substitute and change some of the steps, or to do things out of order. Nonsociable tools are often incapable of handling such changes. Moreover, most activities require the use of multiple tools, multiple technologies, but you would never guess this from the design of the tools: all of them are designed to operate as if they are the only things of interest. Our isolated, context-free intelligent tools cannot be sociable. A sociable design would support the high-level activity as well as the lower-level tasks.

Interruptions

There is another problem: interruptions. Again, most tools assume that no other task interrupts the project and that the activity will be completed in one sitting. But real life provides a continual stream of events. We are continually interrupted by friends, colleagues, bosses. Our personal lives never leave us, so in the midst of unrelated activities, we might still wish to communicate with our friends and family. Many activities take a long time, so we might have to interrupt them for rest, food, or at the end of the day. Finally, people are frequently multitasking. Most activities have numerous simultaneous components, and moreover, we are usually thinking about or doing multiple activities at the same time.

Interruptions produce a heavy mental workload. For example, if we were reading, an interruption means we have to find our place and rebuild our mental structures to resume. If we are involved in deep concentration and mental activity—such as might be required in programming, writing, or design—the disturbance that results from an interruption can be even more extreme. The psychological literature is filled with studies demonstrating the high cognitive workload caused by interruptions and the resulting inefficiency with which tasks get completed. The research literature on the performance of tasks shows that interruptions lead to errors: people forget where they were, sometimes resuming by repeating a task already done or by skipping a step not yet done. Both can have serious negative consequences. In addition, when tasks interrupt one another, each gets done more slowly due to startup time. The total time taken can be far greater than if none of the tasks were interrupted.

Life has a way of making even simple things complicated. In many critical industries, interruptions can be life threatening. To avoid the problems caused by interruptions, pilots in commercial aviation are not permitted to have casual conversations or to interact with the cabin crew during takeoff and landing procedures, the two highest periods of workload in the cockpit. Interruptions during medical procedures are a special concern because in emergency situations, concentration is continually broken through interruptions, many of them requiring immediate action. Even if each of the several tasks and questions being interweaved into the stream of actions is relatively simple, the end result is continual opportunity for errors.

In any emergency situation, new events continually arise, many of which are urgent, causing interruptions to other tasks. As technologies continue to intrude on daily life, the number of interruptions continues to increase, thus complicating even the simplest of tasks, increasing errors, lowering efficiency, and adding to our everyday stress and turmoil.

How can we deal with this issue? When it is not possible to stop the interruptions, then we need assistance to maintain our place among the various activities. Technology needs to be designed with automatic place-saving and reminders built in. It should be designed with the recognition that the worker may leave the activity and, upon returning, will need a quick and easy way to remember just what has been done, what is now required, and what the current status is. All critical information needs to be saved so that even if there is a power loss, it is easy to resume precisely where the interruption occurred. In addition, because a machine being used in one activity might have to be removed so that it can be applied to a different activity, it has to be easy to return it to the original activity and continue where it left off (with the relevant display of state and status). Few of our technologies today support interruptions.

Neglect of Usage Patterns Can Make Simple and Beautiful Things Complicated and Ugly

Appearance matters. Any object is part of its environment, yet when it comes to design, surprisingly little attention is paid to the environmental and social impact of the object. Each piece is designed as if it were an island, independent of actual usage, the surroundings and the people affected by it.

In magazines dealing with architecture and interior design, the photographs of buildings, offices, and homes are always spotless, with nothing out of place. The lawns are carefully manicured. No cracks on the sidewalks. Inside, no papers litter the tabletops, no disarray is visible. In kitchens, there are nice bowls of fruit, but no dirty dishes.

The same disregard for context, for the environments in which things are actually used, is true of the various design contests and magazines for industrial designers. I have been a juror in several design contests where all the creative designs were shown in pristine environments. No wires or plugs, no people, no surrounding activity. I tried to get the rules changed so that in the future, all exhibits had to show the device in use, with all the necessary supporting structure, including power cords, speaker cables, networking connections—everything. My fellow jurors listened patiently, smiling tolerantly; no changes were made.

Much attention is paid to the design of the front of equipment, so that it is made beautiful and elegant, while the back is neglected. But in most business environments, and even in the home, the beautiful front is seen only by the person using the device, while all others—visitors, clients, customers, friends, even family—must look at the rear end. Most rear ends, whether animal or technological, are not known for their beauty. Figure 5.2 shows some typical results, although I am certain that everyone has their own horror stories.

Neglect of usage patterns can transform simple, attractive items into complicated, ugly ones. When the separately simple design components are put together, the result can be infuriatingly complicated. Look at the examples shown in figure 5.2. Not only are they ugly to look at, but the resulting tangle of wires, sometimes in difficult-to-reach locations, complicates the task of connecting, disconnecting, and troubleshooting the connections.

Figure 5.2
Unsociable, unnecessarily complicated appearances. Photographs (a) and (b) were taken at a design conference, this one at the National Science Foundation just outside Washington, D.C., but just look at the mess on the floor between the tables, in (b). A power strip is a simple device, but when used in real contexts, the tangle of wires that result is an eyesore. We had to stare at this mess for two days as we discussed elegance and beauty. Photograph (c) was taken inside a bank in Palo Alto, California. Photograph (d) shows the Engineering Library at Northwestern University, and (e) shows the attractive James Clark Center at Stanford University, but when visitors look in the windows, they see ugly rear ends, as in (f). All these are not only unseemly to look at, but difficult to use: the tangle of wires in difficult-to-reach locations indicates a lack of designing for context of use.

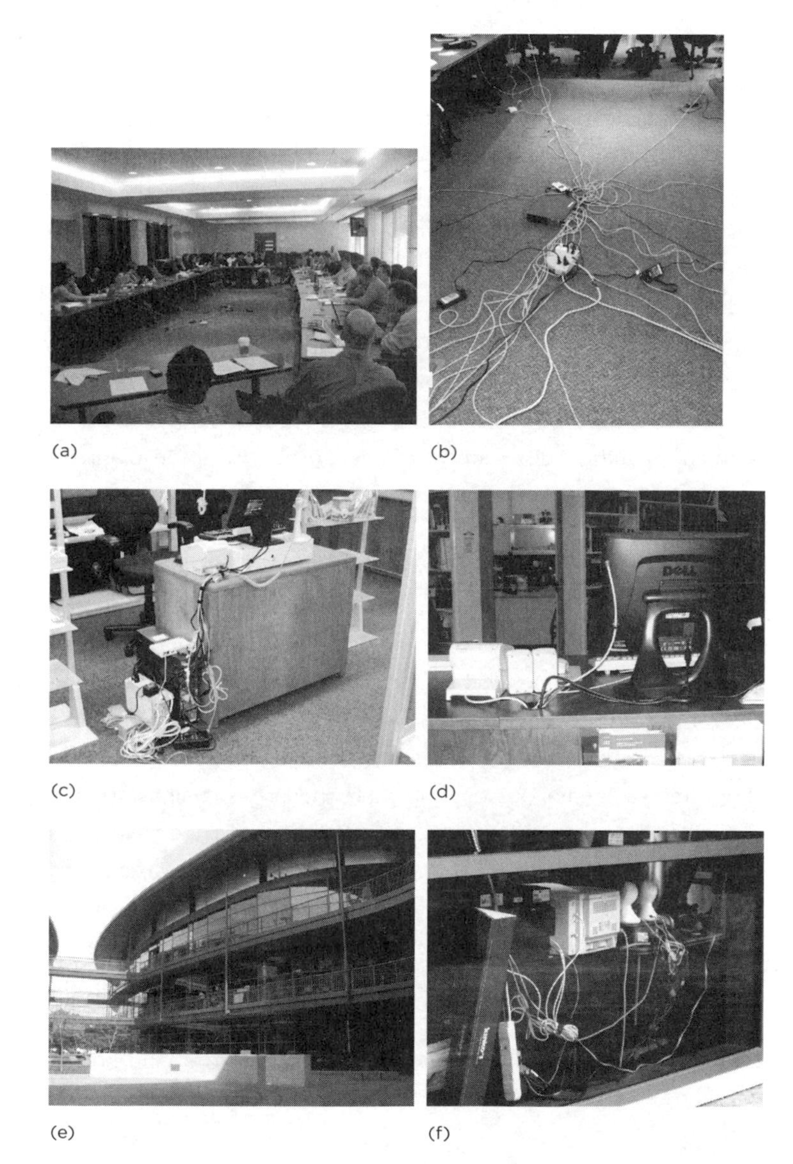
(a)
(b)
(c)
(d)
(e)
(f)

Desire Lines

Walk around parks or college campuses and there, amid the neat sidewalks and pathways, you will find messy trails of people, worn-down dirt paths through the lawns, grasses, and even flowerbeds. The trails are social signifiers, a clear indication that people's desires do not match the vision of the planners. People try to simplify the paths they take when walking, taking short routes rather than long ones, even if it means walking across gardens or scampering up hills (see figure 5.3).

Landscape architects and urban planners are not pleased with the resulting destruction of their grounds. Some planners resent them, treating them as a destruction by thoughtless, lazy people of carefully laid-out plans. These human-made trails are called "desire lines," for they reflect desired paths even though the formal layout of streets and sidewalks do not accommodate them. Wise urban planners should listen to the message underlying these desire lines. When a desire line destroys the pristine plan, it is a sign that the design did not meet human needs.

It is common folklore that some college campuses have had their sidewalks laid out by following these desire lines. This is done by placing no sidewalks until after the buildings have been occupied for a year or so, then laying down the sidewalks along the paths that have resulted from people tramping across the fields between the buildings. I am suspicious of such stories, even though they are frequently told to me. Why? Because it is far too sensible, far too human-centered to have actually been practiced. I haven't ever found direct evidence: only someone who knows someone who heard that story. The practice isn't likely for several

Figure 5.3
“Desire lines.” People try to simplify their lives, preferring short routes to longer ones. When a city provides a rectangular path, such as the gray lines representing sidewalks shown in drawing (a), people will take shortcuts, shown by the dotted lines cutting across the corners. Photograph (b) shows what it looks like in a city. Photograph (c) shows a trail crying out for a sidewalk: the trail is sending a message: “Please, city, put a sidewalk here.” Photograph (d) is an example of how people guide one another to make paths where they seem sensible, even if no official one has been. (Drawing [a] is mine; [b] and [c] are my photographs from Evanston, Illinois; [d] was photographed by Kevin Fox at the University of California, Berkeley: used with permission.)

reasons. When institutions undergo construction, they want to finish and let people get to work. Moreover, if the sidewalks were delayed, allowing a full year of muddy trails, people would complain. Finally, construction budgets, which would include sidewalks, are unlikely to stay around for a full year after the rest of the project's completion. Furniture, by the way, is the same. When a new building is completed for a university, the university is apt to have a one-time furniture grant that it must use by the end of the fiscal year. No matter that some of the space is deliberately unoccupied, designed for future expansion; the furniture must be purchased by the deadline of the budget, with no knowledge of what future occupants might want.

Do sidewalks get built according to desire lines? Perhaps, but to do so is to wage a fight against uncomprehending, very definitely unsociable bureaucracy. Some landscape designers resent people's shortcuts through their aesthetically pleasing layouts. One annoyed landscaper claimed it was "because people, who are notoriously lazy, take the shortest route from A to B rather than stick to pathways that were intended for use." He called these a "blot on the landscape." Don't give in to those pesky people, he argued. Force them to behave: put barriers in their paths:

> To remedy this you need to create a direct pathway or an interruption that is not easily navigated that causes the pedestrian to adopt the intended path.
>
> Obviously an impermeable barrier is one such interruption and that can take the form of a fence, a tree or pond. A border or planting scheme, including pots does the same trick. (Philip Voice's blog, "Landscape Juice")

The irritation of the planner is understandable. Many of us may have shared these sentiments. People cut across lawns and even flower beds. Once a large number of people start using a public space, there is often considerable damage. Building inhabitants paste cardboard or aluminum foil on their windows to block out the glare of the sun, and posters, signs, and notices litter the walls and walkways of college campuses. Yes, these behaviors are irritating, destroying the elegance of the original designs, but if there is any irritation at all, it should be directed toward the lack of sociability that prompts these responses.

Why do people walk across lawns or flower beds? Because the sidewalks and paths are not placed where people need them. Why do building inhabitants block their windows? Because life would be too complicated otherwise: the glare from the sun makes it impossible to do work, makes computer screens unreadable, and heats up the interior. Why are those signs all over the place? Because the signs are needed to understand the complicated world. Signs are often needed to explain how to use things, to explain that something does not work, or simply to announce events because there are no other good spots to put the announcements.

Desire lines tell us how people really behave. Why not use the desire lines as valuable guides and modify the pathways accordingly? Why add unnecessary complication to people's lives?

Do the desire lines indicate laziness? Of course. But laziness is actually a fundamental law of physics, where it is called the "minimization of energy" principle. All physical systems prefer states that minimize energy consumption: people are no different. The landscaper who believes in erecting physical barriers to prevent people from taking efficient paths is not designing for those who must use the space. Unless it is a work of art, a public space is for people. The philosophy underlying human-centered, sociable design is that it is for the benefit of the people who use it, taking into account their true needs and wants.

Although the term "desire lines" originally applied to the trails of people seeking the most efficient path across fields, the term can be broadened to include any indicator of people's natural behavior. One researcher, Carl Myhill, has shown how people's attempts to use badly designed systems leave trails behind that are the equivalent of desire lines. Skid marks on the road, wear marks on benches and stairs, even forms filled out with the proper information but not in the squares intended by the designers. Myhill shows that if one simply watches people's behavior, the discrepancies between the designers' intentions and observed behavior provides valuable design information.

Desire lines are important signifiers of desired behavior. Wise designers and planners pay attention to these signifiers and respond appropriately. A relatively simple way to simplify things is to use the trails left behind by people's actual behavior to design systems that support their desires.

Sometimes, however, design is meant to provoke. This is where it is an artistic work, intentionally created to stimulate speculation and discussion, perhaps to provoke debate. In this case, people's complaints are perhaps best ignored. For example, I own the juicer designed by the French designer Philippe Starck (he calls it *Juicy Salif*). It's not very good at making juice and there are numerous reviews posted on the Internet about its ineffectiveness for this purpose. So what? This is meant to be art: Starck himself has said that the purpose was to create conversation. I treat it as art and place it on display in my living room. I use a more functional juicer in the kitchen.

Sometimes it is appropriate to ignore people's wishes and to force behavior to fit the vision. This is true for art, meant to create debate, but it is also true in cases where the goal is to prevent unsafe, dangerous, or illegal behavior. In these cases, it is reasonable to put up barriers against "inappropriate behavior," to put up warning signs or even to pass laws punishing the behavior. Here is where deliberately making the undesirable behavior difficult to do should be the goal of the designer. Signs don't always work, as vividly illustrated in figure 4.6 (p. 103). Sometimes added complexity and difficulty are called for. Desire lines indicate true preferences, but not all preferences should be accommodated.

Traces and Networks

Desire lines are visible in the physical world because when people wander across fields they disturb the ground, leaving marks and injuring plants. The more that other people follow the same paths, the more intense the marks, the greater the impact on the ground and plant life. The same holds true of all physical actions, for each leaves some impact of use. When people read books, their trail can be followed by the dirt marks and smudges, turned corners of pages, folds, and written annotations. Even the book spine reflects usage, opening automatically to a frequently used section.

In the world of electronics, we also leave trails, except that these are not visible without the aid of technology. Even the simplest of activities is apt to leave a trail. Walk through a hallway and a video camera records the passage. Use a credit card and leave behind a record of what was purchased, for how much, and where you were when you did it. Look up some information electronically, and you leave behind a record not only of what was asked, but what activities immediately preceded and followed the question. Messages, whether by voice or electronic messaging services, must be recorded in order to be delivered to the recipient, but even after delivery, although both sender and recipient may attempt to delete or destroy them, their traces stay behind.

The traces we leave behind provide valuable information, not only about our own behavior but about general human behavior. Today there is a growing body of scientists devoted to studying these resulting networks of interconnections: people connecting to people, to physical locations, to systems and organizations. These traces can be used to simplify our lives or to complicate them.

One of the more powerful uses of trails as social signifiers is the records left behind as people read material in magazines, books, scientific journals, and, of course, on the Internet. The importance of these traces has been recognized since the early 1900s, probably starting with Paul Otlet's *Traité de documentation* in 1934 and Vannevar Bush's "memex" concept of 1945.

Otlet did his work in Europe in the middle of the twentieth century, and as a result, the Second World War interfered with and minimized the impact it might otherwise have had. Bush, an electrical engineer who helped lead the American scientific effort during the war, was more influential. In an article published in the popular magazine the *Atlantic Monthly* in 1945, Bush advocated a "Memory Extender," a memex, that would display books, films, and other reading material and automatically create and follow cross references from one to another (remember, this was a half-century before the invention of hypertext and the Internet). Bush recognized that the trails formed by readers would be valuable in their own right. They would simplify the efforts of scholars trying to do research on a topic. Thus, wrote Bush:

> Wholly new forms of encyclopedias will appear, ready-made with a mesh of associative trails running through them, ready to be dropped into the memex and there amplified. The lawyer has at his touch the associated opinions and decisions of his whole experience, and of the experience of friends and authorities. The patent attorney has on call the millions of issued patents, with familiar trails to every point of his client's interest. The physician, puzzled by its patient's reactions, strikes the trail established in studying an earlier similar case, and runs rapidly through analogous case histories, with side references to the classics for the pertinent anatomy and histology. The chemist, struggling with the synthesis of an organic compound, has all the chemical literature before him in his laboratory, with trails following the analogies of compounds, and side trails to their physical and chemical behavior.
>
> The historian, with a vast chronological account of a people, parallels it with a skip trail which stops only at the salient items, and can follow at any time contemporary trails which lead him all over civilization at a particular epoch. There is a new profession of trailblazers, those who find delight in the task of establishing useful trails through the enormous mass of the common record. The inheritance from the master becomes, not only his additions to the world's record, but for his disciples the entire scaffolding by which they were erected.

In the early days of bibliographic reference, back when we believed all behavior was cooperative and benign, it was thought that these wandering trails would be of great value to others. Thus, both Otlet in the period 1910–1934 and Bush in 1945 envisioned a world where readers left behind the trails they followed that would be as valuable as the individual books, for they would allow a novice in a topic to follow in the footsteps—the trails—of masters, following the same connections they had made. Today's realization of many of these ideas still languishes. The Internet allows for the linking of ideas, but only through links that are explicitly placed by human Web site developers (or machine algorithms), or by links formed through the use of search engines. Both Otlet and Bush had in mind that a reader might follow someone else's implicit steps.

Following the trails of other researchers sounds like a wonderful idea, but I am not convinced it has much value. Would it truly simplify our work, or would all the false trails and restarts simply complicate our lives? How would we know which paths would be valuable for our purposes? For example, suppose you had followed my trail of investigations in writing this section of the book. You, the reader, could follow my wandering through the *Britannica Online Encyclopaedia*, Wikipedia, the Web site for *Boxes and Arrows*, the Web site for the School of Information Management and Systems at the University of California, Berkeley, plus email messages on this topic between me and members of the research community. Along the way I discovered new people and new sources of information, but I also hit a lot of dead ends, talked with a number of people only to find that they were unable to help. If you, the reader, were to follow these trails, you might not be enlightened. After all, this particular path took me almost two months, with a lot of false movements and floundering. In many cases, it would be best to follow not the path but only the summary of that journey.

There is one other problem with the blind following of trails left behind by others: they may be false trails, deliberately placed to fool or mislead us. The early workers in hyperspace and the World Wide Web all assumed benevolent users, people who only intended to guide and aid others. Today we know better. Many people are out simply to have fun; many are out just to cause trouble. Many have narrow views of the world that they want to force on others; they strive to propagate their own belief structures while eradicating all hints of other beliefs.

Scholars leave trails of their work through the references and citations in their papers. You can see an example of this in the notes of this book: ideas that were derived from other people's works are attributed to those other people. The legal system is one of the pioneers of compilations of citations, lists of the references from one legal opinion to another, starting in the late 1800s. Lawyers have long recognized the importance of these trails, especially as much of law is based on precedents, so it is important to know which cases cite which other cases. In the 1950s, Eugene Garfield recognized that it would be valuable to do a reverse analysis of scientific papers and see how many studies that followed a given publication referred back to it. So was born modern citation analysis, at first done by hand, but today completely automated. The citation index is not only useful for research scholars who are able to trace the impact of a scholarly idea forward in time through the citations, it has also turned into a widely used rating tool for the importance of a scholar: "and how many people have referenced your work in the past year?" a dean is likely to ask of someone during an evaluation for hire, retention, or promotion.

The social signifiers formed by all the physical and electronic traces of our activities can be valuable supplements to our lives. Social networks are important ways of linking people to one another, including friends, common interests, and communities of education, work, and play. The resulting linkages of people provide valuable insights into the interlocking of people's interests and communities, make it easy for people to keep in touch with old acquaintances and discover new ones, make it possible to get questions answered and receive advice, and also provide a rich bed of information for thieves and law enforcement agencies, advertising agencies and salespeople, former friends and troublemakers, whether for your benefit or distinctly opposed to it.

Today, these social signifiers form the basis of an important tool in the world of electronic information: recommender systems.

Recommender Systems

Why do we tend to look at books that are listed on a bestseller list? Or, when in an electronics store confronted with a daunting array of similar looking devices, why do we often find the clerk offering to help by pointing to one of the items and saying, "This is our most popular item"?

These are examples of primitive recommender systems, with the recommendations based on sheer popularity of the items, very much like selecting among unknown restaurants by shunning the empty ones. Popularity is not a bad starting point; after all, if everyone likes something, it must be OK. These recommender systems differ from the expert recommendations given by a reviewer of books, electronic devices, or restaurants and published in a newspaper or magazine. In these professional reviews, we have to decide whether we agree with the biases of the reviewer, who often seems to be writing for the benefit of other reviewers, not the average reader, user, or restaurant customer.

Bestseller lists are based on sales figures, which is both their virtue and their weakness. Nobody is average: everyone is distinctive in some way. How much better if the recommendations would be from people whose interests, biases, and skill levels were about the same as our own? That's what modern recommender systems are. Because people's behavior is captured electronically, whether by the usage of the computers, telephones, or credit cards, it is possible to segregate users by a wide range of characteristics, including the activities they are doing, age, place of residence or work, and prior interest in related items.

In the virtual world of information spaces, every activity leaves behind traces. Search queries spell out one's interests. So too do the pages that have been read, especially those that are referred back to. In stores, the items browsed and purchased provide a record of interests, much as footsteps in the snow provide a record of travels. The difference is that recommender systems can select among the trails, following only those left by people whose interests and goals are similar to yours.

When you buy an item, the store knows what you have purchased, and if it is a virtual store, it knows all the items you considered but did not buy. When you watch a show, video, or movie, the provider can determine just what parts you viewed, what you skipped, and what parts you repeated. So too with electronic books and articles. The details of your activities are available: not just what you watched, read, or did, but how, when, and sometimes even with whom.

Recommender systems now proliferate. Booksellers tell you what other people whose interests are similar to yours have liked. Same with the purchase or rental of products and services. Same with music and sports, restaurants and clothing. The same principles can be used by law enforcement agencies, creating a sophisticated version of profiling: "people like this one," a system might tell the police, "have caused problems," "rob banks," "murder," or even "complain about police actions."

Do these systems simplify our lives, or do they complicate them? They work by making general assumptions about people based on common backgrounds and interests. They work because they amass a huge amount of information about individual people. They work well on average, but not in every particular case. We benefit when the system recommends a book or restaurant, but leaves us free to ignore the recommendation. In these cases, the systems simplify our interactions with the complexities of life. But when the systems fail, when they are used to predict behavior of individuals, especially when they are trying to predict deviant and unlawful behavior, the use of average predictions is inappropriate, and both the likelihood and cost of errors are high. In these cases, false predictions create complication for individuals and society.

Support for Groups

Support for groups is the hallmark of sociable technology. Group support is obvious in the activities shown in figure 5.4, but groups are almost always involved in activities, even when the other people are not visible. All design has a social component. Groups can complicate activities that have different philosophies, views, and agendas.

In the case of the conference photographs shown in figure 5.4, the design support has as much to do with conference structure as with physical space. The conference is deliberately located in an isolated camp in the Rocky Mountains, in the middle of winter, so it is not easy for the participants to go anywhere. It is organized with considerable "empty" time, which means that they naturally congregate and discuss the conference topics. The room itself is sociable, with food, chairs, and places to gather, as well as a deliberate group activity: the jigsaw puzzle of figure 5.4b. The goal of these spaces is to facilitate interaction, making possible the informal discussions and debates that are so critical to the diffusion and advancement of scientific ideas across different groups.

Notice that this conference has optimized social interaction through isolation. The conference encourages focused interaction, restricted to the participants. Here is a case where less is more: less opportunity provides more focus and depth.

People are naturally social and communicative. By appropriate sociable design, we can take advantage of people's skills, enlist them in understanding the activities taking place so that if difficulties arise, the possible courses of action are understood. Understanding is what transforms complex systems into simple ones.

(a)

(b)

Figure 5.4
Social groupings. People work well in groups, whether in informal conversations, as in (a), or when trying to solve a problem, as in (b). Photographs from the annual Human-Computer Interaction Consortium, held in the middle of the Rocky Mountains, Colorado, USA.

Group understanding is often more powerful and robust than individual understanding.

Design of both machines and services should be thought of as a social activity, one where there is as much concern paid to the social nature of the interaction as to the successful completion of the activity. That is sociable design.

Figure 6.1
Services are like matryoshka dolls. In a typical service encounter, customer and employee face each other across a counter (a). To the customer, the employee represents the service, but each employee has to deal with services internal to the company. Services are like these Russian nested dolls, where just as each doll has yet another one inside it, each service contains other services within it (b and c).

6
Systems and Services

Most of my work has been with computer and telecommunication companies and with startup firms that make use of these technologies. These companies manufacture electronic products: computers, cameras, cell phones, navigation systems, and so on. In the early days of these new technologies, people had enormous difficulties understanding and using them. These were interactive devices, where an action by a person would lead to a change of state of the machine and then the requirement to do some new action. In many cases the person and the device had to engage in a form of conversation in order to set up the right parameters for the action that was to take place. As a result of the difficulties being faced, computer scientists, psychologists and other social scientists, and designers developed a new discipline, interaction design, to figure out the most appropriate ways of handling the interactions. As the technologies have evolved, and as the sophistication of the people using them has increased, the field of interaction design has had to deal with more and more advanced techniques and philosophies of interaction. From understanding and usability the field expanded to incorporate emotional factors, toward a focus on experience and enjoyment. Today, more and more products contain hidden, embedded microprocessors (computers) and communication chips. As a result, interaction design is now a major component of almost all design.

The world of services is different from that of products, in part because they have not been studied as much as products. Although one would think that service providers should also adhere to the standard themes of good interaction design, that is, good feedback along with coherent conceptual models, in practice it is not so simple. Services are often complex systems, barely understood even by the service provider, with multiple components spread across many geographical locations and divisions of the company. This creates huge barriers to the development of good models for the operation and makes feedback especially difficult.

At first glance services and products appear to be distinct entities, but the attempt to define either is surprisingly difficult. A service is often defined as a helpful action or work done for someone else. Often the only difference between a service and a product is point of view. In some sense, every product provides a service to its user. Both a camera and a refrigerator seem like pretty typical products, but they provide valuable services to their owners. The refrigerator maintains food at safe storage temperatures: this is a service. A camera is a physical product, but what it offers its owners is a way of remembering and sharing experiences, which is a service.

In similar fashion, to the company that manufactures the automatic teller machines (ATMs) used by banks, these are products. The bank provides the product for use by its customers.

But to the customers, the ATM provides a service, making it more convenient for them to do their basic bank transactions.

Services are often exceedingly complex, for many of our most common services—home utilities, telephone service, government services such as licenses, passports, and income taxes—have immense bureaucratic rules and regulations, hordes of back-office people, and often multiple divisions of a company all having some say in responses to any question that is not routine. As individuals, all we see is the front end of the service, the visible part exemplified by the person, mailing address, telephone contact, or Web site that is our source of interaction.

All that stuff behind the scenes—those mysterious operations that give rise either to smooth, efficient operations or to confusing, mindless ones—is called the "backstage." The "frontstage" or "onstage" components are the parts visible to the customer, for example, the bank clerk waiting to help you. Backstage refers to all the activities that occur out of sight of the customer, for example, the backroom operations of the bank, done either in offices not visible to the customer or, most likely, in a completely different building, perhaps far away from the physical location of the bank. Many of the bank's backstage operations are not even done by the banks, but by the various entities of the international banking network, which include companies, consortiums, and governments. The backstage operations are essential to the appropriate completion of the service transaction, but the customer is often only aware of the visible frontstage.

The distinction between front- and backstage components of a service implies a neat separation, but this is misleading. Everything has a front and back, so each of the backstage components has its own front and back. Part of the bank that is backstage to the customer is frontstage to the clerk, and what is backstage to one clerk is frontstage to yet other clerks, who in turn have their own front- and backstages.

Services are recursive. They are somewhat like matryoshka dolls of figure 6.1, the Russian nested dolls: when you open one up, it holds another very similar doll inside, and if you open that one up, there is yet another doll. The design of modern systems and services must cope with this recursiveness, with the fact that what is to be designed depends on the point of view. Do you take the point of view of the customer, the clerk, the backroom assistant, or the central administration? Answer: you must consider them all.

The internal backstages of services are critically important. This is the operations side, where all the work gets done. If operations fail, or if they are clumsily executed, the product or service fails. Complex products have complex operations and technologies behind them, which means that many people work behind the scenes to make possible the smooth, effortless operations experienced by the people interacting with the out-facing component. But these people have their own tools, each of which has an in-facing and an out-facing component.

Successful products have to incorporate all the various layers of in-facing and out-facing components, harmoniously supporting all the visible and hidden services and operations. Products exist within a complex web of interactions.

The design problem here is immense: how can all the participants involved, both customers and employees, receive the information they need so that they can understand the operations?

Feedback and conceptual models are most important at two times during usage. One is when the product or service is first experienced, for now these aid in learning what to do and what to expect. The other is when there are problems or unexpected delays. Perhaps more information is needed, or approvals, or something is going wrong somewhere in the chain of operations. With products, it is usually relatively easy to handle these situations. But with services, especially the complex ones that involve multiple organizations and locations, it is very difficult to provide the proper information. Service design is far more complex than product design.

Services as Systems

Many services are both social and complex systems. Many are provided by large organizations, with components in very different geographic locations. Quite often the different parts of the organization do not understand or communicate well with one another. And many services involve different organizations, and communication among them is particularly difficult.

It is easy to find examples of the complexity of services: think of almost any interaction with a governmental agency. There are many potential sources of difficulties, from the interaction with government employees, the complex set of rules and regulations that must be followed, the complexities of the forms that must be filled out, and then to the impenetrable delays that occur as the request moves from one office to another, perhaps from one agency to another. Even if everyone is helpful and friendly, the sheer complexity of the operation coupled with the relatively poor interfaces among all the components can lead to frustrating experiences.

The only way to solve the complexities of services is to treat them as systems, to design the entire experience as a whole. If each piece is designed in isolation, the end result may be of separate pieces that do not mesh well together.

Consider some examples.

The Acela Express Amtrak Train

David Kelley, one of the three cofounders of the major American design firm IDEO, told me with great pride how his company had approached a request from Amtrak, the American passenger train service, to redesign the interior of its railroad cars. Amtrak wanted to bring out a new high-speed line, Acela Express, traveling from Washington, D.C., northward along the eastern coast of the United States to Boston. Amtrak asked design firms to submit proposals for redesigning the interiors of the trains to attract more riders.

IDEO's response was to say "no." That is typical of design companies, by the way. IDEO was practicing what designers call "design thinking," which means, among other things, to start by first determining what the real problem is. I often explain it this way: Never solve the problem the client has asked you to solve. Why? Because the client is usually responding to the symptoms. The first job of the designer, sometimes the hardest part of the entire task, is to discover what the underlying problem is, what problem really needs to be solved. We call this finding the root cause.

In the case of train service, because many of the regular riders complained about the experience, Amtrak assumed that this meant that the interiors of the trains should be redesigned. This description tries to solve the symptom, not the cause. The proper solution requires a systems approach, not just the redesign of one of the many parts, such as the train interiors. Amtrak, to its credit, agreed with this analysis and allowed a complete reconceptualization of the entire service experience, a task that IDEO was happy to work on.

IDEO and its partners, Oppenheimer and Company Brand Consultants (O+CO) and Steelcase, recommended that Amtrak treat the travel experience as an integrated system, starting with the decision to travel by train rather than airplane or car and then continuing on through all the stages of the trip: purchasing the ticket, the experience at the station, both on departure and arrival, and the experience on the train. They identified ten steps to train service:

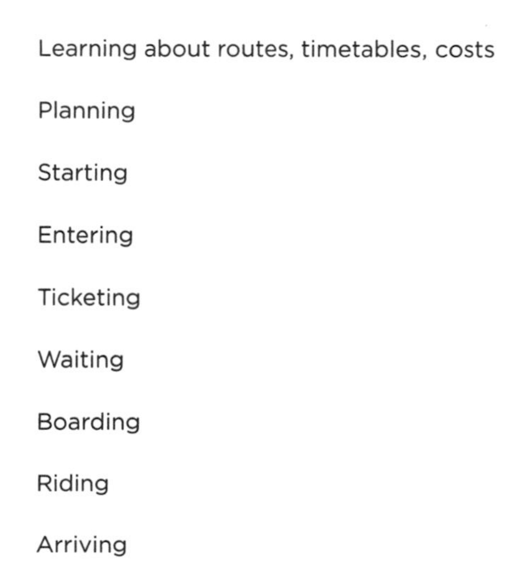

Learning about routes, timetables, costs

Planning

Starting

Entering

Ticketing

Waiting

Boarding

Riding

Arriving

Continuing the journey (train terminals are seldom the final destination for a traveler).

Each of the ten steps was considered a design opportunity: each was essential to the success of the whole. Note that the original design request was just for one step, the riding experience, which is only one part of the entire system. IDEO and its design partners wisely redesigned the entire system, from Web sites to waiting rooms to the interior of the passenger and dining cars. They redesigned information kiosks at the train terminals and even the uniforms of the staff. The design team was multidisciplinary, including experts in human factors, environments, industrial design, and branding. The result was a very successful transformation of the train experience. The redesign increased the number of riders and established the route as the most popular in the entire United States.

Apple's iPod Music Service

Portable music players are a very popular product. Carry along a small device with hundreds or thousands of your favorite musical selections, available to the listener whenever and wherever desired. Ever since the first portable music players were developed using cassette tape recorders in the 1970s, portable music players have revolutionized the way people listen to music. The first major success was the Sony Walkman, released in 1979.

It was the computer revolution, however, with the advent of tiny processors and huge-capacity memory systems along with Internet commerce and compression systems that reduced the size of recorded music files, which set the stage for the next revolution in the 1990s. Now the electronic players were much smaller and easier to carry than the Walkman. Moreover, each could hold thousands of songs, something never before possible. The first barrier to the success of these devices was legal ambiguity of getting the music. Although it is legally permissible to purchase music and then to make a copy for one's own personal music player, it is not permissible to distribute that music. The second barrier was the complicated set of steps required to get music into the device: copying, compressing, and transferring the music to the player was a daunting task for the average person.

When Apple entered the market with its products, it created a revolution in the distribution of music. Apple soon took over, not only dominating the sale of digital music players, but also changing the way the music companies thought of their products. Everyone believes that Apple took over the music-player business through its superior design of its device, the iPod, introduced in 2001. No, the iPod, although a truly excellent product, is not the secret to Apple's success. The secret is that they understood that the core problem was not just the design of the product: it was to simplify the entire system of finding, buying, getting, and playing music, and also to overcome the legal issues. Note that at the time, a number of companies were already selling digital music players, some of which were quite attractive and functional. But these were isolated products. Most music could not be acquired legally for use in these devices. Getting the music into one's computer and then into the player required manual operations that were more complex than the average person was able or willing to do. The fundamental problem was the integration of all the parts into one seamless experience. Like the train experience, this system has multiple stages:

Getting licensing agreements from music producers (making it legal to get the music)

Browsing the music store to find the desired music

Purchasing

Transfer 1: getting the music into one's personal computer

Transfer 2: getting the music into the music player

Synchronization and sharing of music libraries

Listening to the music

A digital rights management system (DRM)

Encouraging other companies to manufacture add-on devices, such as external speakers

Control over the retail environment

A trademark and licensing scheme

Apple treated iPod as a service, not as an isolated product. They therefore worked diligently to ensure that all stages were handled seamlessly, resulting in an excellent customer experience. In quick summary, Apple was the first company to negotiate the legal licensing of music at a reasonable price per song. Second, they designed a Web site and a companion application for the computer that made browsing through music, searching, and experimenting with new performers fun and enjoyable. Third, Apple made the purchasing trivial, and the downloading of purchased music into one's computer effortless.

Apple also designed the iPod system so that when the iPod is plugged into the computer, the transfer of files into the iPod occurs effortlessly and easily. Finally, Apple's design of its music player, the iPod, was excellent: it is easy to listen to music on the computer, and to stream it among networked computers or even to home audio and television systems.

At the time the service started, music sellers were terrified that people would freely transfer music to one another without payment, so they insisted on digital rights management systems that would prevent this. Apple complied, but it restricted the licensing of its DRM so that the music it sold could only be played on Apple devices, ensuring what marketing people call "lock in": The larger the collection of songs purchased from Apple, the more the person was locked in to continual use of Apple products. The extensive library of music could not be played on devices made by other companies (unless they had licensed the DRM rights from Apple, which was an infrequent event). The digital rights management issue is still an enduring issue, not just for music, but for all media, such as films, video, and books. Media companies are exploring various options that protect ownership rights but are not as restrictive as some of the early implementations. Apple has also loosened the restrictions.

Finally, Apple developed an ecosystem, encouraging other companies to develop add-on products such as loudspeaker systems, accessories for playing the music over automobile sound systems, accessories that enhanced the iPod's capabilities, turning it into a stopwatch, a voice-recording device, and a storage device. All of these were licensed by Apple, which received commissions (royalties) on sales: a risk-free source of revenue.

Apple treated the entire effort as a seamless system. Even the design of the box in which the physical devices were packaged was exemplary. Many companies try to save money on packaging: Apple spent extra money, treating the package as yet another chance to offer the customer an engaging, delightful experience. Apple understands that the user experience starts with opening the box, which should be just as exciting and pleasurable as the rest of the experience.

The story improved with time as Apple expanded the range of devices in their product portfolio to include mobile phones, portable computers and display pads, and other devices that interconnect computers, telephones, cameras, video, and sound systems. Although the repertoire goes far beyond music to devices that manage photographs, videos, movies, games, newspapers, magazines, books, and other media, all follow the system design point of view. The physical structure, the capabilities, and the names of the devices have changed several times, but the general philosophy of making the entire system seamless and effortless has endured. As business conditions change, Apple keeps changing the offering, but it still excels at three things:

Creating cohesive systems, not isolated products

Recognizing that the system is only as good as its weakest link

Designing for the total experience

System thinking: that is the secret to success with services. Whether it is a Disney theme park, an Apple service, Netflix movie service, FedEx or UPS's delivery services, or purchasing from Amazon: because these companies design the entire system, as a customer order or request goes through the backstage operations, these companies keep informing the customer of its progress every step along the way, always estimating shipping and delivery times, allowing the customer to modify the orders far into the process, and making sure that the entire experience is handled well from the customer's point of view. The operations behind the scene are smooth and efficient—this is the province of the operations staff, often using sophisticated mathematical and computer simulation tools to ensure optimal efficiency. Even the dull, mundane operation of shipping a package from one destination to another can be transformed into a positive experience by appropriate attention to keeping the customers informed. Good system design considers the entire process as a human-centered, sociable system.

Service Blueprints

Services are complex systems, with many interactions taking place. Not only are they difficult for a person interacting with the service to understand, they are difficult even for the service provider to understand. Service designers have grappled with this problem of complexity, attempting to develop a means of diagramming all the encounters. In the early 1980s, Lynn Shostack, then a senior vice president of Bankers Trust Company, proposed a method of showing both the time course and the depth of the interaction through a process she called "service blueprinting." I'll demonstrate how it works by borrowing from the more recent publications of a research scientist at IBM, Susan Spraragen, figure 6.2.

In addition to the two dimensions of time sequence and depth of operation, Susan Spraragen has adapted the blueprint to capture the emotional state of the customer, as shown in figure 6.3. By adding the customer's reactions to the service blueprint, Spraragen's diagrams show the full impact of the experience. She calls these diagrams "expressive service blueprints."

All of these attempts at describing the customer's experience are important tools in developing the proper service structure, but none of them captures the entire complexity. We need to cover the emotional states of the worker as well as those of the customers. We need to show the backstage operations in much more detail. And none of these diagrams indicates how we would explain to the customer (or for that matter, the company staff) what is actually going on.

Still, the blueprint might provide the proper model. Perhaps a simplified blueprint could be presented to customers and workers, showing for each transaction exactly what stages have been done and where the process is at the moment, and indicating what future steps remain to be taken.

A good customer experience is not difficult to provide when everything works smoothly. But when both the requests and the service organizations are complex, things are apt to go wrong. Perhaps the information is incomplete, perhaps a key person is absent, or perhaps the system has to await parts, authorization, or another critical item. The difficult part of design is to ensure

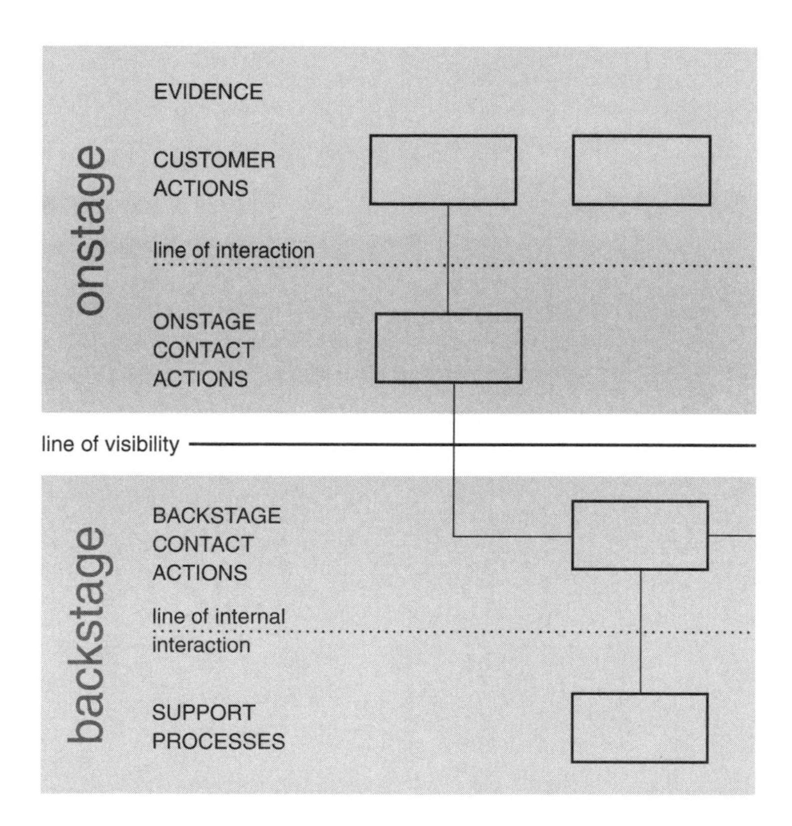

Figure 6.2
The basic service blueprint. The horizontal "line of visibility" separates frontstage (called "onstage" in this figure) from backstage operations. The vertical dimension represents all the parts of the organization (in the backstage) that are involved plus all the components relevant to the customer (onstage). The horizontal axis represents the sequence of stages the process goes through, with time flowing from left to right. From Spraragen and Chan 2009.

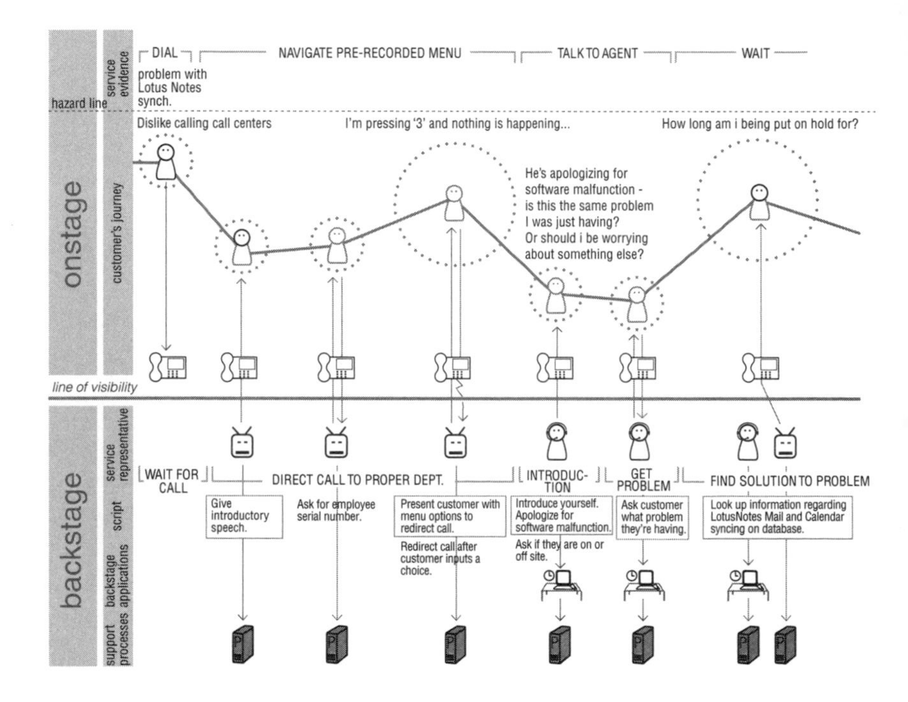

Figure 6.3

An expressive service blueprint. This scenario shows a customer calling a service desk for help with a computer application. The bubbles around the icons that represent the customer indicate the level of frustration: large bubbles indicate high frustration. (Notice the last step in this figure where the customer is told to wait and is put on hold.) The vertical position of the customer on the page indicates the customer's comfort level with the service provider: when the distance to the line of visibility is small, the customer feels close. Finally, the thoughts of the customer are indicated by the text. From Spraragen and Chan 2009.

that things work well in the face of unexpected difficulties. Until we provide feedback and conceptual models, we will fail to provide an excellent customer experience.

Designing the Experience

"If you go to a good hotel and ask for something, you get it. . . . If you go to a great hotel, you don't even have to ask." The Ritz-Carlton Hotel chain's philosophy—they want to be great hotels.
—Paul Hemp (2002)

To the individual customer or employee, a service is often an experience. This means that as much attention has to be given to the well-being and comfort of the employees as the customers. One of the most interesting examples of this observation is over a century old, illustrating that the secrets to effective service design lie with the management of people, not with technology.

In the late 1800s and early 1900s, when transcontinental passenger service across the United States had just begun, Fred Harvey opened a chain of restaurants across the western United States. The goal was to provide restaurant and hotel service for the passengers of transcontinental trains. Trains made regular stops in order to restock the steam engines with coal and water and to allow the passengers to get out and stretch. Harvey realized that this provided an opportunity for a restaurant business, with restaurants located at the train stops, working efficiently enough to feed a trainload of passengers in the short time permitted for their stop. Although Harvey did measure the time it took his employees to do their work, he insisted on attention to the details of how his service people interacted with the customers. His empire lasted around seventy-five years, with sixty-five restaurants between Chicago and San Francisco, serving over fifteen million meals a year. The secret of this success story was both in scrupulous attention to detail and also in the care and training of the personnel.

Today, the Harvey chain's level of attention to detail and to the interactions between staff and customers can still be found in hotel chains and restaurants. Paul Hemp, an editor of the *Harvard Business Review*, spent a week training to be a room-service waiter at a Ritz Carlton hotel in Boston. The training emphasized empathy with the guests and anticipation of their needs. The care and concern shown by the employees has to be genuine: the staff really has to care about their guests. Every morning, they get together to discuss the day's guests and their desires. They review the service philosophy of the Ritz every day, even if they have been working there for decades. Their job, the staff is told, is "to make guests feel good so they come back."

"No one remembers if you served a guest from the left or the right," says John Collins, the hotel's human resources director. "But they do know and remember if the service is genuine, if you actually enjoy being of service. They can tell if you're forcing a smile." Staff at the Ritz are empowered to make decisions, to override rules where necessary for the guest, to bring guests extra items, and always to be prepared for unexpected requests. If a single guest has ordered coffee or wine, think ahead: might there be someone else in the room? Carry an extra cup or wine glass. Hemp writes that he thought these little attentions excessive until he once delivered a single cup with a coffee order, only to discover two people in the room.

This kind of exemplary service cannot be faked: the staff has to believe in it. This means that as much attention has to be given to the needs of the staff as to the guests. The staff is treated well, and the continual reviews, the emphasis on helping one another, and the ability to take independent action when required to assist a guest gives each of them a sense of pride in working there. As the staff members are trained to develop empathy and to provide a pleasurable experience for guests, they end up taking considerable pleasure in their achievements.

The Ritz is a very expensive hotel. My business students always immediately critique the study as reflecting the niche luxury market. A normal company, they say, could not afford the extra cost of such detailed attention to the welfare of both staff and customers. I disagree.

Web sites count as services, and so the very same lessons apply. Some Web sites recognize repeat visitors, making useful suggestions, but in a nonintrusive way. Web sites have a number of special features that make them different from personal, physical interactions. There may be millions of people using the site, with diverse needs and understandings of the offerings. Somehow, the site has to be able to cater to everyone, but without diminishing the experience. A real test of a company is how it responds to criticism, especially when it has taken some action that irritates a lot of customers. This is where the true test of sociability arises: when things go wrong.

Some inexpensive hotels attend to their customers in ways that are very much appreciated yet do not entail added cost. For example, the Club Quarters chain of hotels caters to business people (whose companies must be a member: Northwestern University provides me with access). Prices are low and service is minimal. For example, there is usually only one person in the downstairs lobby: customers check themselves into the hotel by inserting their credit card into a machine and receiving their room key. Customers check out in the same way: no human interaction is required. This is advertised as a feature: "instant check in/check out." Room service is also minimal, but each room has a list of nearby restaurants that will deliver food to the room. And a closet on each floor is well stocked with extra coffee, soap, shampoo, and the common things guests might need: anyone is free to make use of it. Finally, the hotels are centrally located in major business cities across the United States (and in London), and provide free Internet access, a good desk, light, and electric outlets that are desired by the business traveler. All this self-service frees up the one attendant in the lobby to provide helpful advice and deal with any problems. No frills, no fancy service, but for the busy businessperson who usually has no need or time for these expensive hotel amenities, Club Quarters shows that care and attention to the customer can be provided without high cost.

As with all services, there is always a temptation to add new features, adding to the number of options that can be offered to customers, but at the cost of increasing complexity. At Netflix, a movie rental service with a particularly outstanding Web site, they decided to allow their customers to have multiple lists of preferences. Each customer has a queue of films waiting to be viewed. Netflix works by allowing only a fixed number of films to be held by a customer at once, but for unlimited time. Whenever a movie is returned, the next one in the queue is mailed. Each customer also has a profile, stored internally by Netflix, that specifies their particular likes and dislikes (after viewing a movie, the customer is asked to rate it). Netflix realized that many customer accounts are used by several people such as family members and roommates, so they added a feature of allowing one account to have multiple profiles. After a while, Netflix decided that this added more complexity than value, so they announced that they would discontinue the service, stating:

> Please know that the motivation is solely driven by keeping our service as simple and as easy to use as possible. Too many members found the feature difficult to understand and cumbersome, having to consistently log in and out of the website.

To Netflix's great surprise, many customers objected, both directly to Netflix and on many discussion groups throughout the Internet. Eleven days later the company recanted:

> Because of an ongoing desire to make our website easier to use, we believed taking a feature away that is only used by a very small minority would help us improve the site for everyone. Listening to our members, we realized that users of this feature often describe it as an essential part of their Netflix experience. Simplicity is only one virtue and it can certainly be outweighed by utility.

Customers were elated. One put it this way:

> My bummed-out feeling that I had toward Netflix last week has been utterly replaced with a general feeling of satisfaction and goodwill for a company that respects its users enough to listen to their needs at this level. Thank you, Netflix!

Numerous studies have shown the importance of effective recovery after a mistake. Some studies show that a company that corrects failures properly may be liked even more than companies that never make a mistake. This result is controversial and some more recent, carefully controlled studies seem not to confirm this finding. Nonetheless, all studies show that companies that admit their mistakes and immediately correct them benefit more than those that try to hide or deny their mistakes. A mistake, whether it is a product that doesn't work correctly or, as with Netflix, a decision later regretted, gives the company an opportunity to demonstrate how much it cares about its customers, how well it listens, and how sincerely it corrects its errors. Service is about experiences. Actions matter; but sincerity, honesty, and personal attention can also have a large impact.

Creating a Pleasant Out-facing Experience: The Washington Mutual Bank

Washington Mutual, a bank that used to have branches across the United States, patented the floor layout and design of their banking offices (figure 6.4). Although some claim the patent silly, just another example of what ails the patent system, my interest has nothing to do with its legal status but rather what it indicates

Figure 6.4
Washington Mutual's (patented) bank design. Not your ordinary bank: Individual "islands" for interaction with bank employees. Wires nicely hidden. No waiting lines. Friendly interiors. A special play area for children.

about the company's priorities: to improve the banking experiences of its customers. Regardless of the validity of the patent, it is clear that Washington Mutual understood the importance of the customer experience.

Figure 6.4 shows the office that was located near my home in Evanston, Illinois. No long counters that act as barriers between customer and bank staff. No long waiting lines and unfriendly interiors. Instead, customers are greeted when they enter the bank by a concierge who directs them to the island where a bank representative will assist them. A children's play area is nearby (visible in the upper center part of figure 6.4).

Notice the way the bank staff and the customers interact. Rather than have a counter or desk, with the two people separated by a barrier, here both stand together, jointly viewing the transaction. In a traditional bank, all seems secretive. The bank staff utilize hidden screens to review the customer's account, and the information is hidden from the customer's view. In the stations of figure 6.4, both view the same screen together. This makes a very distinct change in the entire experience, for now, instead of facing a nameless bureaucracy, with hidden, secret data, which is how the average bank sometimes feels, both staff and customer enjoy a feeling of cooperative, sociable interaction.

Unfortunately, the Washington Mutual Bank no longer exists. It collapsed during the economic crisis of 2009 and was acquired by JPMorgan Chase Bank, who announced that they would not use the floor design. Why? Washington Mutual aimed their services at individual investors. For these customers, the personal, informal attention provided within the novel bank design was appropriate and extremely effective: business thrived. Chase Bank aims primarily at business customers and large accounts, wealthy individuals to whom it offers private banking services. The design of the Washington Mutual Bank does not allow for the private, confidential discussions required for these customers. So, back to the traditional bank layout, with bank staff isolated from customers by counters and bulletproof glass. The retreat to convention may be appropriate for the business needs of Chase Bank, but it is an unfortunate retreat from an innovative, effective sociable design for us ordinary bank customers. Other banks, however, are introducing similar designs.

It is essential that services be designed with both the staff and customers in mind. Happy employees make for enthusiastic, courteous interactions with customers. The Washington Mutual Bank understood the importance of this. When I talked with the bank staff, they were unanimous in their enthusiasm for the bank layout. They showed me each area and illustrated how it worked. They asked me to withdraw $1 from my account just so they could demonstrate how money was handled in a safe, efficient manner, automatically dispensed from machines upon each request. This made it easy for the customers, but would make it very difficult for a thief. The staff were quite insistent on showing me every detail, and it was clear that they took great pride in their workplace. When I asked for permission to photograph it, they had to huddle for a moment to decide (there are obvious security risks from potential thieves who might take photographs), but it was also clear that they were proud to work in a place that an author (which is how I described myself) might wish to write about. All in all, the bank seemed to treat both staff and customers well. The results showed in the way they interacted with customers.

Washington Mutual transformed the routine, administrative interaction with a bank into a sociable experience: service design can be made sociable.

Service Design as Factory Design

Consider what has been learned from the design of factories. An efficient factory operation requires, among other things, the management of inventory and bottlenecks. What is inventory? It is a buffer. A queue of items waiting to be worked on. In the store, the inventory is the items waiting to be purchased. What is a bottleneck? The name comes from the design of a bottle with a narrow neck, the size automatically limiting the amount of liquid that can emerge at one time. In a factory, a bottleneck is anyplace where the flow of goods is limited, so that material piles up behind the bottleneck, creating a queue. Bottlenecks are usually caused by a lack of resources: in the bottle, it happens because the neck is (purposefully) made narrow. In the factory, a bottleneck is caused by a machine or group of workers not keeping up with the pace of activities. It is when cashiers cannot keep up with the number of customers, or when there aren't sufficient taxicabs for all the people waiting for rides, or when the size of the highway isn't large enough to carry the demand.

Many bottlenecks are easy to understand, but not necessarily easy to fix. Normally, eliminating bottlenecks requires more resources. But adding more resources can be expensive, especially if it requires adding more people, equipment, or in the case of a highway, more traffic lanes. A considerable amount of effort by management experts and engineers is spent devising means of reducing bottlenecks and improving efficiency. Much of the work done within the science of operations management is devoted to this topic. Some problems can be solved by simple rearrangements of the processes. Some can be minimized by developing more efficient procedures. But some are more complex.

Traditional service design is aimed at increasing efficiency and lowering costs. Consider the waiting rooms of a health care clinic. A waiting room is a queue. Queues of patients are desirable from the point of view of the hospital administrator. After all, the hospital must pay the salaries of their physicians and staff; it must pay for the upkeep of expensive equipment. With waiting rooms filled with patients, the X-ray equipment or MRI scanner are never sitting idle. Physicians, nurses, and technical staff never have to waste valuable time waiting for a patient. But from the patient's point of view, it would be best to have waiting rooms filled with physicians. What about emergency rooms? Here, time is critical, so the balance shifts in favor of the patient. In my studies of hospital operations, I once saw an emergency room that essentially was a waiting room filled with medical personnel: physicians and staff were chatting, checking their email, and catching up on personal affairs while awaiting new emergencies. This is a rare occurrence, however. The physicians said that this was an extremely unusual situation for them.

There are many types of factory floors. One extreme is the job shop, where everything is made to order. The other is the assembly line, where a steady flow of repetitive operations is performed. In the job shop, machines are organized by their type: all the milling machines here, the presses there. In an assembly line, the machines are organized to make the flow of the product more efficient, so that each machine is positioned to accept the output of adjacent machines. A stamping machine might be adjacent to a welder, which might be adjacent to a bolting assembly. Errors are very few, quality high, because when an error does occur, the root cause can quickly be found and eliminated.

Job shops can't arrange their machines according to work-flow because each job is different, each apt to require very different procedures than the one preceding it. Because of the continual change in requirements, there are high setup and cleanup times, and even the work itself is slow, usually manually performed, and special for each case. In an assembly line, just the opposite holds: low setup times, efficient actions, low or even no cleanup. A job shop might have a high error rate with variable quality. When each job is different, lessons learned from one may not apply to another.

Many services are job shops. Restaurants, for example, are primarily organized like job shops: each order might be very different from the one before, and although the variety is restricted by the menu, the number of combinations and permutations is still enormous. Fast food restaurants, on the other hand, often model themselves after the assembly line, with each food item being carefully measured, proscribed, and operated upon in a consistent manner, using as much automation as possible. The hospital is a job shop. Even the building is organized like a job shop, organized by the type of procedures to be done. Laboratory tests are done in one section of the hospital, X-rays and scanning in another. Hospital wards are also segregated by the type of procedures required. Each medical specialty is located in a different part of the building, even if they must work together as a team. The patient is shuttled back and forth. High setup costs, and manual, slow actions (subject to frequent errors).

Businesses seek profitability, which is sensible, for the business that continually loses money will not last, no matter how good the product or service. Modern management tends to focus on numerical measurement, arguing that it is only through measurement that progress can be monitored and maintained.

If you can't measure it, you can't improve it, so goes the belief among many in both science and management. Measurement has been a powerful tool for improving efficiency, but it is wise not to let the power of measurement override analysis of the important issues. When it comes to people, not everything we believe to be important can yet be measured. On the other hand, much that we know is unimportant is easy to measure. Unfortunately, the need to measure often dominates the need to consider the important variables. In the educational system, it is easy to give tests and devise numerical or letter assessment of students. The relationship of those grades to life achievement is negligible. Even so, we persist in numerical assessment.

Numbers dominate. Scientists measure what they can measure and pronounce the rest to be unimportant. The most important parts of life are qualitative, but still we persist in measuring and recording. I refer to modern medicine and, in particular, the hospital. In medicine, the situation has reached the point where there is so much to measure, so much to record, that no time is left over for the patient.

Hospital Care

It's 6:30 in the morning. I'm with a group of surprisingly wide-awake, cheery physicians and nurses, doing grand rounds on the pediatric care ward of one of the best hospitals in the United States. I'm part of a study group for the National Academies, looking at the ways in which information technology is used in health care. This hospital is a leader; I see computers everywhere.

I've been spending a lot of time in hospitals recently. Not as a patient, as an observer—following doctors and nurses on their rounds, watching patients get admitted, nurses doing shift changes, pharmacists filling prescriptions (in these big hospitals, they fill millions each year, all the while trying to eliminate drug interactions and errors), and then watching nurses deliver the prescribed medication to their patients, waving barcode readers over the prescriptions, the medication, and the patient's armbands.

We walk down the hall toward the first set of patients. We are quite a crowd: the attending physician and five medical residents, physicians completing the last stage of their training, plus one or two nurses and then the several members of the study team of which I am a part. The attending physician is both responsible for the patients' treatments and also supervising the residents. Each of the residents is wheeling a computer cart in front of them. Many places call them "COWs," for computer on wheels, but one hospital explained that they had switched the name to WOW, workstation on wheels, after a patient who heard the physicians conferring outside her room talk about "the COW" thought they were referring to her. A COW is a chest-high cart, with computer screen and keyboard at a height appropriate for reading and typing for a standing person, with the computer itself and batteries located at the bottom of the unit. Five COWs, plus a nurse wheeling a big filing cabinet of papers, plus the rest of us. We took up a lot of space. Whenever we stopped at a patient's doorway to review progress, the residents would flip through the windows displayed on their computer screens and summarize status: "calcium level is fine, white count low." Each resident had a different piece of the patient, or to be more precise, had screens that described the test results of different laboratories.

The patient was a bunch of numbers. Moreover, the numbers were not organized by symptoms or diagnoses: they were organized by what tests were run and which laboratory within the hospital had processed the data. Current results were in a different place from the patient's history. Different hospitals might have different laboratories, so their results would be organized differently. But the attending and resident physicians and nurses were experts at piecing together a mental model of the state of the patient from all these numbers.

A hospital is a complex place, where multiple operations must interact smoothly with one another. Problems arise at interfaces—any interface, be it person and machine, person and person, organizational unit and organizational unit. Consider the problem of finding a bed for a patient, another activity that I observed with great interest in a room filled with people and computer screens, along with paper charts. Somehow I had assumed that when patients were admitted, or released from the emergency, intensive care, or delivery wards, or from the operating or recovery room (the hospital is filled with all sorts of specialized places for patients), they would simply be sent off to any available bed. But no: beds are specialized by the kinds of services available there and several other considerations. As a result, we have yet another interface, yet another source of complexity.

In the study team I was on, looking at the role of computers in medicine, one physician told us that she is only allowed fifteen minutes to attend to each patient in her internal medicine practice, but it can take as long as twenty minutes just to fill out all the information required by their medical information computer system. She has to force herself to look at and interact with the real patient. The Vanderbilt University Medical Center estimates that nurses only spend one-third of their time in direct care of a patient. Half of the remaining two-thirds of their time is spent on documentation and medication record keeping.

Where Is the Patient?

"That's interesting," I said to myself, stepping into a room filled with displays. There were numerous infusion pumps, computer readouts, and monitors. The entire room was filled with the red glowing lights of display readouts and the dim white of graphs on computer screens. "Interesting," I said, "you have brought all of these monitors into one place so you can see how all the patients are doing."

"No," said one of the physicians, "what do you mean?"

"So where are the patients?" I asked, expected to be told that they were in adjacent rooms.

"Right there," said the physician, obviously puzzled by my request. "Right there in the room, right in front of you."

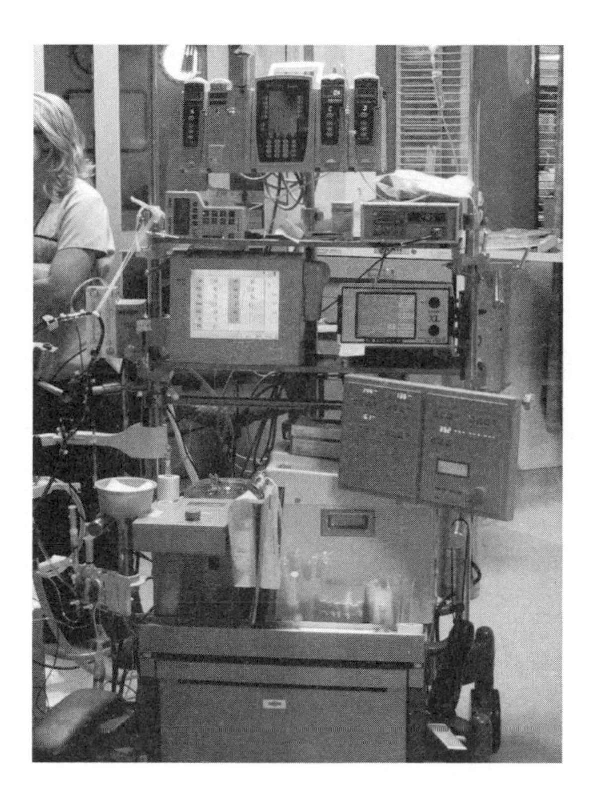

Figure 6.5
Hospitals are dominated by measuring devices. A typical patient's room, dominated by instruments, many with incompatible readouts and controls, many of which must have their measurements recorded manually, all leading to medical records dense with numbers, hiding the condition of the patient as a person.

I looked closely and still couldn't see a patient. One of the nurses walked over and pointed. "Oh," I said.

There were so many medical devices, so many readouts and displays, that I could not even see the patient. Now this was an infant ward, so this particular patient was tiny, but even so this is a good illustration of modern medicine: from the point of view of the physicians, the patient is a set of test results and numerical readouts. The patient as a person tends to be forgotten.

I saw this later in a different hospital in yet another ward. The attending physician would stand outside of the patient's door and listen to the review of the test results by all the residents. They would then discuss the results and make further recommendations. Then, as we all left to go to the next doorway and the next patient, the attending physician would knock on the open door, stick his head in and say, "How are you doing today, Mr. Forbes?" That was the extent of patient interaction.

Where is the patient in all of this? Forgotten. Not only that, all the measuring equipment could very well be harmful. Consider this abstract from the medical journal *Neonatal Network*:

> The neonatal intensive care unit (NICU) environment is often very noisy because of such factors as the frequent use of ventilators and other mechanical equipment, the use of monitor alarms, and staff conversations and traffic. For premature infants, who are unable to tune out normal background noise, exposure to high levels of sound can cause stress and may disrupt their normal brain development. . . .
>
> These results confirm that NICUs tend to be noisier than recommended and indicate that even changes to the NICU environment

thought to be helpful or innocuous may need to be evaluated for their effect on overall sound levels.

Patients are numbers, digital readouts, and test results. All these tests come at a price: concentration on the tests instead of the patient, overcrowding of rooms with equipment, often of different manufacturers, using different styles of interaction, inviting mistakes (which might then be blamed on the nurse rather than on the multiple design flaws and incompatibilities), and a resulting noise level that can be harmful to the patient. Is medicine getting better? Yes, but how much harm is it doing?

The State of Service Design

It is said that the Japanese first eat with their eyes and then with their mouth.
How a meal looks is as important as how it tastes.
—Japanese folk sayings

The total experience we have with a product goes far beyond the product itself. It's all about expectations: how they are set up, and then, of course, how well they are met. It's about the way we anticipate, use, and then reflect on our experiences. It's about the image the product imparts to those who own and use it, about the image of the company that makes the product, and about the services surrounding the product, from initial advertising, through the experience of selection and purchase, delivery and initial installation. It's also about usage, continued usage, and interactions with the company for service, maintenance, and upgrade. In other words, it covers every aspect of interaction, from the initial engagement, to the experience, to how well the company maintains the relationship.

Although the design of products has received considerable attention, the study of service design is still in its infancy. As a result, less is known about the design of services than products. Moreover, product design is sexy: it is easy to hold design contests where companies submit fancy photographs of their products and then harried jurors spend a few days trying to select the prize winners. I have served my time as juror in a number of these contests, and even though the jurors would prefer to judge all aspects of the product, it is impossible. Far too many products, far too little time. There is no way a jury can evaluate the feasibility or practicality of the hundreds of entries (in one contest I judged, there were thousands of entries). As a result, these prizes primarily reflect appearance and say very little about how well the things function, or if they are even usable at all, to say nothing of whether or not they are likely to succeed in the marketplace. Product appearance is where the action is: that's what controls the prizes.

Services don't have the glamour of products. In many cases, there is nothing to see. Service design is about procedures—which means they have to be analyzed in action. Less money is invested in the study of services than in products, even though many companies succeed or fail on the quality of their services. This is a worldwide problem. The Köln International School of Design in Cologne, Germany, says that the result of this lack of attention is a poorly functioning system:

> disfunctionality and formlessness are not unusual in this sector: endless waits, broken appointments, unfriendliness, unreliability as well as the torture of formalities that seem absurd determine the everyday service from the customer's point of view. And the suppliers of service moan about the customer's lack of price willingness, about unreliable loading factors and unmotivated service employees. (Description of the program in Service Design from the Web site of the Köln International School of Design, Cologne, Germany)

In Germany the amount of money invested in research and development for manufacturing is about thirty times as much per employee as the amount invested in services. Although the quotation is from Germany, these problems are worldwide.

The poor standards for services reflect several causes. First: managers often take services for granted, and do not pay much attention to the design of the service side of an organization or to the training of staff. Second, in the modern, numerically obsessed management culture, the emphasis is on driving down costs and improving whatever aspects of performance and efficiency can be measured. But services deal with people, where the things that can be measured usually deal with duration and number of operations, not with customer or worker satisfaction.

The importance of services is well known, with many published studies and books. Quick reviews of the *American Marketing Association* journals (e.g., *Journal of Marketing*), the *Journal of Operations Management*, the *MIT Sloan Management Review*, or the *Harvard Business Review* show that studies of services are frequent. However, in business schools, there are few studies of the human side of service design. Most studies of services have focused on operational efficiency and, in particular, on mathematical modeling to optimize the expenses of dealing with the expected customer load. As a result, the experience side of service lacks design principles for either those being served or the staff who do the serving.

Yes, services are complex. But services are intended to help people. Moreover, they are staffed by people. In the rush for modernization and productivity, we have tended to overlook the value of the human experience. Measurement of critical variables is good, for it allows us to focus on weaknesses and the direction of change. We must not forget the human element, however. We need to reduce the complexity, we need to make the interactions less complicated. In the rush for efficiency through measurement, we should not forget the wisdom of Albert Einstein, the physicist, who is reported to have said "not everything that can be counted counts, and not everything that counts can be counted."

Figure 7.1

Waiting may be a necessary part of life, but that doesn't mean we enjoy it. Waiting at airports. The top picture was taken at Chicago's O'Hare airport, the bottom two at Cancun, Mexico.

7
The Design of Waits

A waiting line is a simple phenomenon, but even so, it can give rise to considerable complications, along with the resulting confusions, frustrations, and buildup of emotions. Unexplained waits are annoying; unfair waits can be anger inducing. A wait is always a sign of a bottleneck in processing, a place where there is more demand than can be accommodated. Waits are side effects of complex systems.

There will be waits whenever one system has to send items or information to another. It doesn't matter whether the interaction is between two organizations, two people, two machines, or a person and a machine or organization. If the receiving system is ready first, it must wait until the next item arrives. If it is ready last, earlier arriving items will be waiting to be dealt with. What happens with the items while they wait? There must be some place to hold them.

When more people arrive at a location than can be handled at once, there has to be some way of handling them. If they line up in a row, we call this a queue or a line. If they mill about, we might call it a crowd or a mob. In a computer system, we keep the waiting items in buffers. In a store, the items on the shelves waiting to be purchased are called stock or inventory. In a hospital, patients are put into waiting rooms. Waits are everywhere, once you start looking for them: books on a bookshelf, food in the pantry, and any items that are stockpiled. There is an entire science that deals with methods for handling queues, buffers, and inventory. The management concept known as "lean manufacturing" is specifically designed to minimize the amount of stock and inventory being held.

The Psychology of Waiting Lines

Even though waiting lines are conceptually simple, they complicate our lives enormously. The person in line soon develops a long list of questions about efficiency, fairness, and even the nature of the line itself.

When there are multiple lines it can be difficult to determine which line does what. After getting in a line, the lack of feedback causes anxiety: how much longer will this take? Will I miss my next appointment? What if I finally get to the end only to be told that I was in the wrong line, or that I didn't have some necessary paperwork? Why are the other lines moving more quickly than my line? Why is it that some people seem to get special privileges or manage to barge ahead of everyone else? Why is it so inefficient?

If lines are truly inescapable, what can be done to make them less painful? Although there is a good deal of practical knowledge shared between corporate managers, very little has been published about the topic. The classic treatment is provided in David Maister's "The Psychology of Waiting Lines." Maister suggested several principles for increasing the pleasantness of waiting but much has been learned since 1985. This chapter follows the spirit of Maister's original publication, but with considerable revision in light of more recent findings.

Waiting lines are well studied in the field of operations, but the emphasis is on the mathematics of efficiency: what is the best scheme for handling customers while keeping costs as low as possible? How many clerks are needed to handle the expected number of customers? Such computations are necessary, but they leave out the human element: what the experience is like for both customer and clerk. My primary concern is with the experience.

How do we enhance the experience of waiting lines? It is a design question, with a set of design principles as the answer. I suggest six design principles based on recent research in the behavioral and cognitive sciences.

Six Design Principles for Waiting Lines

1. Provide a conceptual model.
2. Make the wait seem appropriate.
3. Meet or exceed expectations.
4. Keep people occupied.
5. Be fair.
6. End strong, start strong.

1 Provide a Conceptual Model

Perhaps the most critical of all design components is the conceptual model of the experience. Conceptual models can transform confusing products and services into coherent and understandable ones. They can do the same for lines.

It is essential that the environment provide clear and unambiguous indication of what each line is for, how to enter the lines, and what information or material is going to be required once the front of the line is reached. Clear and unambiguous social signifiers are essential here. This requires all the skills of a good designer: good observational studies, good ideas, good prototypes, and continual observations, checks, and refinement.

A good conceptual model sets up expectations and aids in understanding of the actions that are taking place. For the model to be effective, there must be ample feedback. Uncertainty is a prime cause of emotional irritation: a good model coupled with proper feedback removes this source of anxiety. When problems arise, people need assurance: they need to know what is happening. Even the explanation that the source of difficulty is unknown can be reassuring: it indicates that the relevant people are aware of the problem and working on it. The goal is to minimize uncertainty by providing reassurances and evidence of care.

One of the worst places in which to wait is a hospital. Anxious patients and family wait in limbo, often in dull, dreary surroundings that help set a tone of negative anxiety. What is going to happen? How serious is it? How long must we sit in this room? Is there anyone who can give us any information? Usually the answers to all of these questions are "we don't know; nobody knows."

There are many legitimate reasons for the delays and for the lack of information. Quite often it truly is the case that nobody knows. There are also operational and legal reasons for withholding information, including simple overload on the part of the hospital staff. But a major reason is lack of thought and appropriate design. Hospitals are designed with many concerns in mind: the insurance companies, the owners, the administration. And the physicians, nurses, and staff. And, yes, the patients. Waiting rooms for friends and relatives? Yes, they are needed, so they are added. But it is the rare hospital that spends time, effort, and money to enhance these periods of uncertainty for patients, relatives, and friends.

The task isn't easy: hospital personnel are extremely busy. There are intense emotions involved. There are concerns about the appropriate way to deliver and explain the complexity and uncertainty of medical conditions. Medical information and records are also subjected to privacy restrictions on what can be told to others, and in many cases, medical personnel overinterpret the restrictions for fear of legal consequences. It is not an easy situation to be in, not an easy situation to design for. But clearly, the experience could be much improved.

2 Make the Wait Seem Appropriate

When people must endure waits, they should know why. Moreover, they should agree that the wait is unavoidable and, therefore, that it is reasonable that they have to wait. This is the role of feedback and explanation and is an important component of fairness (principle 5). The reasonableness depends on the situation.

Here is where the conceptual model is so critical. If people have a good understanding of the backstage actions taking place, they are apt to accept waiting as necessary and appropriate. Without a good conceptual model, people will make one up, and invented models are very apt to be inaccurate and badly misleading.

If the wait is caused by factors outside of everyone's control, as when an airplane flight is delayed because of severe weather, then the reason for the wait is understandable and accepted. This doesn't mean the wait will be well tolerated: the other rules still apply, but at least one barrier is overcome. When there is a clear reason for a wait, such as a busy restaurant, or a filled amusement park, the wait can be tolerated as long as its duration is appropriate to the reason. When there is no apparent reason, or worse, when the reason is visible and seemingly inappropriate, the wait will not always be tolerated. If service in lines is slow, but all the clerks are clearly working hard and all the positions filled, the wait might be tolerated with patience, as in customs and immigration lines at airports. But when there is a large crowd awaiting service, yet only a few people providing it, then the tolerance shifts to blaming the service provider for not reacting to the need. Worse is when there are service people available but not helping, especially if they seem to be relaxing and enjoying themselves. Whenever service people take their breaks, they should be out of sight of the customers.

Note that this perception of appropriateness meshes with the conceptual model: people want to know why the wait, why the workers do not appear to be working, what is happening. The perception of appropriateness ultimately derives from a combination of information about the situation and the conceptual model. The wait must seem appropriate, both in its cause and its duration. Similarly, the service provider should be perceived as responding appropriately to the demand.

3 Meet or Exceed Expectations

Experiences should exceed expectations. Many places try to give time estimates of the expected wait. Experience shows that the time should always be overestimated: if the actual waiting time is shorter than the expected time, people are likely to be pleasantly surprised.

Providing relevant activities for the people in line helps transform the otherwise dreary waiting into a positive experience. The goal is for the person to exit smiling, saying "that wasn't bad," or perhaps even truly enjoying the experience. The fact that people often start off with negative expectations about a line can actually help because this makes it easier to find something that enhances their perceptions of the wait.

4 Keep People Occupied

To understand this rule it is important to keep in mind the difference between physical variables and psychological ones. They are not at all the same, even though we may use the same names to describe them. Thus, physical time and distance can be precisely specified and measured, but people's perceptions of distance and time are governed by psychology, not physics. Moreover, there is a significant difference between the immediate perception of duration and distance and the later memory of them. Psychological duration is dramatically affected by mental activity. Thus, a period filled with events feels as if it is passing more quickly than the same physical time without events (an empty period). These distinctions between filled and empty time or space can be used to advantage in the design of waiting lines. Keep the lines moving fast, make them look short, keep them filled with interesting things to look at and interesting activities to do.

One trick to making a line enjoyable is to make it seem as if it isn't a line. Good examples can be found in the entertainment industry, especially in theme parks. Parks operated by Disney are famous for how they handle lines, curving them around so that they are visually short, providing entertainers to engage the people in line, ensuring they are enjoying themselves. Moreover, long lines can be made to look short by clever placement of the route, with bends hiding part of the line ahead. In some cases, components of the goal activity can be moved forward, which has the effect of making the line shorter. In a restaurant, people can start off in a bar area where they can be served drinks and appetizers. In organizations, the necessary paperwork can be prepared while waiting. Educational material can be presented. And as I discuss later in this chapter in the section entitled "Double Buffering," entertainment sites can create briefing rooms and other activities to occupy people who are waiting their turn to enter a limited capacity ride or activity. The activities not only help people to endure the wait, but by providing relevant activities early, they actually shorten the waiting period.

5 Be Fair

Emotion is heavily influenced by perceived causal agents. If the wait seems reasonable, with nobody to blame, it will not necessarily trigger a strong negative emotion. The emotion comes when there is something to blame, even if it isn't true. Thus, if the line appears to be arbitrary, unpredictable, and worst of all, unfair, emotions rise.

Do others have an unfair advantage? Do other people cut in line? Are there special people who don't have to wait in line? All of these can lead to heightened negative emotional states, far more severe than the state of having to wait longer than expected. One of the strongest determiners of a good experience is whether treatment was fair. With long waiting lines, resentment builds when some people appear to take advantage and get ahead of others. There are multiple places where fairness has an impact on the situation.

One problem with multiple lines is that the other line always appears to be moving faster. This is true of cars in highway lanes and customers in shopping market checkout lanes. Whatever lane you switch to, the other one seems to move faster. The perception occurs because the amount of time the service provider takes to process a person varies. Some people are processed quickly, others incredibly slowly. And no matter which line you are in, it always seems as if it is indeed the slowest. We note and remember when people in other lines start moving faster than the line we are in; we tend not to notice when our line moves quickly ahead of the others. It is this asymmetry that leads to the perception of unfair lines. Psychological experiments have demonstrated that even when all lines are moving at the same average speed, people perceive whichever line they are in as moving most slowly. This is yet another reason why the best line design is to use a single line branching off to multiple servers at the end: with only one line, the perception of fairness increases. And with multiple servers, the line moves more quickly than multiple lines, each with a single server.

6 End Strong, Start Strong

What parts of an event do we remember? This is a much-studied question by psychologists. Unique experiences always stand out. But if everything is relatively homogeneous (such as the act of waiting in line, from entering through leaving), then the most important influence on memory is the ending, the beginning, and the middle, in that order. This is called the "serial position effect." Some studies have even demonstrated the counterintuitive result that a long, unpleasant event can be perceived more positively if at the end, an extra period is added with a less unpleasant (but still unpleasant) component. This is counterintuitive because the longer event has all the unpleasantness of the shorter one, plus even more, except at a lower level of unpleasantness. But it is the memory of the ending that dominates. The moral of this experimental finding is clear: always end on a positive note.

Design Solutions for Waiting

Different cultures have different expectations of the waiting experience. One major difference is whether there should be a line at all. Polite, orderly queues are the rule in some cultures. In others, people try to force themselves to the front, with the noisiest or most forceful winning. Travel around the world and you will find the differences are striking: orderly queues of patient people in London; disorderly mobs clamoring for train tickets in Beijing and Casablanca. In much of Asia people will crowd around counters, each person demanding the attention of the service providers. Although many Westerners are appalled, the system works well. A Chinese friend explained that with a typical, orderly (Western) line, people wait for a long time with nothing happening. In the apparent disorder of the Eastern crowd clustered around the service agents, people can get attention almost immediately. Although the agent's attention to their problem is quickly interrupted by the demands of other people, a tiny amount of the transaction gets accomplished. In the end, both systems may take equally long, but in the Asian method there is a continual feeling of progress.

In the strange, artificial world of communication between machines, a polite form of mob culture is frequently used. When one machine wishes to send a message to another using Ethernet, the standard way computers on a local network communicate with one another, all the machines use the same channel. How does each know when it is its turn to communicate? Over the years, many different systems have been tried, but the most popular one today is very much like the system used at counters in Asia.

Each machine breaks up its message into a number of relatively small packets. The machine watches the single, shared channel and as soon as it sees a break in the message stream, it tries to send a packet. If some other machine also tries at the same time, the two packets collide, and both machines have to quit and try again. Of course, this would cause the same collision once again, so the rules built into the machines require them to wait for some randomly determined time before trying again. If another collision occurs, then the wait time is increased. The more collisions, the longer the wait before being permitted to try again. The system works remarkably well and is part of the international standard for Ethernet machine-to-machine communication.

Note that the scheme of getting access to message sending means that, just as in the mobs of people clamoring for attention, there is continual progress. Each machine is slowly getting its message sent. The more people clamoring at the same time, the longer it takes for the entire message to get through. But because everyone follows the same rules, the system works very well. No lines are required, no handing out of numbers. There is no central controlling authority.

Where does such a structureless system work with people? It seems to work in many cultures around the world, not just in Asia. It also works in normal conversation, as long as the number of people in the group is not too large. Each speaker normally waits for a lull in conversation in order to speak, and if two people speak at the same time, they usually quickly work out which person gets to continue. One sees a variation of this when multiple traffic lanes converge: in many societies, drivers alternate turns, making the convergence simple and efficient. In some cultures, it is a free-for-all, with each driver trying to squeeze into any perceptible space; this can result in a complete jam, where nobody can move.

To someone accustomed to lines, the mob system seems more complex, less fair, than requiring orderly lines. To people used to more immediate gratification, the use of orderly lines may be conceptually simpler, but it delays progress.

The rules for appropriate queuing behavior across cultures get quite complex. In some places, it is permissible to let others into the line, either just in front or in back of you, without consulting the people behind, all of whom suffer. In other places, this is socially frowned upon. Can one person reserve a space for another? In a long line, is it possible to leave temporarily and then regain the same space? Often the answer is yes, but only if permission has been requested and granted from the person behind. What about selling your place in line, or hiring others to stand in line for you? These behaviors are common when people must wait overnight or longer, which can happen when tickets are being released for some highly popular event, or some new, exciting consumer product is being released.

Cultures can be changed. McDonald's changed queuing behavior in Hong Kong:

> The social atmosphere in colonial Hong Kong of the 1960s was anything but genteel. Cashing a check, boarding a bus, or buying a train ticket required brute force. When McDonald's opened in 1975, customers crowded around the cash registers, shouting orders and waving money over the heads of people in front of them. McDonald's responded by introducing queue monitors—young women who channeled customers into orderly lines. Queuing subsequently became a hallmark of Hong Kong's cosmopolitan, middle-class culture. Older residents credit McDonald's for introducing the queue, a critical element in this social transition.

So, yes, culture can be changed, but don't count on it. And even if it is changeable, it is apt to take years, perhaps decades. Of all things changeable, culture is the hardest. Moreover, the same *Encyclopaedia Britannica* article from which the McDonald's quotation was taken also warns:

> It would be a mistake, however, to assume that . . . innovations have an identical, homogenizing effect wherever they appear. . . . It remains difficult to argue that the globalization of technologies is making the world everywhere the same. The "sameness" hypothesis is only sustainable if one ignores the internal meanings that people assign to cultural innovations.

One Line or Many? One- and Two-sided Cashier Stations

Consider the checkout lane of a cafeteria. In one typical arrangement, a cashier sits at a cash register and tallies up the cost of the individual items of the customer, who pays and leaves. Both cashier and customer spend a good deal of time waiting for the other, which is not efficient. Where are the inefficiencies? In the setup and cleanup times. Here is a typical description of the cashier and customer experience:

> Wait in line until the previous customer has left and the clerk seems ready
>
> Walk up to the cash register
>
> Put items on the counter

Wait while the clerk tallies up the items and presents the cost

Find credit card, money, or check

Make payment

Put away any change, receipts, and credit card, wallet, or checkbook

Gather up purchases

Walk away from the cash register

Now the next customer must go through the same steps, starting with a delay to determine whether the cashier is truly available. The cashier spends a lot of time waiting for customers to show up, to unload, to pay, and to clean up so the next customer can arrive. In turn, customers wait during the inefficiencies of previous customers, and then extend the wait through their own inefficiencies.

From the customer's point of view, the process is wait, move up, position the items properly, wait, pay, pack up, and leave. From the cashier's point of view, it is wait, tally the items, wait, receive the payment, provide receipts, and wait. This task analysis, which identifies the activities of all participants, is helpful in assessing the particular problem areas where changes can be made. What are some ways to improve the single line checkout inefficiencies? One way is to reduce the time taken for these operations. Another is to allow the cashier to service other customers while the first customer sets up and, later, cleans up. Yet another approach is to provide buffering space for the setup and cleanup activities to take place without interfering with the previous or next customer. Let us look at some of the known solutions.

Double Buffering

In the world of computer graphics, where it is very important to be able to display images rapidly and smoothly, one standard technique is to alternate between two different storage areas: two buffers. While one buffer is in use, the other buffer is being filled. Then, when the display of the first buffer is done, the display switches to the second buffer, so there is no interruption in displaying the image. And while the second buffer is now being used to display the image, the first buffer is being filled with the information required for the next image.

The very same procedure can be applied to amusement parks, or any situation where people are served in batches. Consider shows and other amusements that take a batch of people all at once. While the first batch of people is enjoying the experience, there is a line of waiting people. How do we make being in that line enjoyable? We turn it into its own experience.

We take a second group of people equal in number to the amount that can take part in the experience, and usher them into a special place for "briefing" or "preparation." There, the waiting people are entertained, perhaps by having the mission they will be engaged in explained, or perhaps by being told the story and background information of the event they are waiting for. The result is that people perceive this as part of the total experience, rather than as waiting in a line. Yes, there still are other people behind them waiting in line, but that line is shorter by virtue of the fact that twice the number of people as before are enjoying the entertainment. Everyone wins.

Spatial Double Buffering: Two-sided Checkout Lanes

A form of the double buffering principle can be seen in the design of a two-sided cash register. Here, the cashier is in front of a cash register with customers on both sides—the right and left. The cashier waits on the left-side customer and, when finished, turns to the right-side customer who is ready and waiting to be served. This gives the left-side customer time to pack up and leave and for the next customer on the left to get ready to be served. The cashier alternates between serving the two sides, giving the customers on the side not being serviced time to set up and get ready at the start of the transaction and time to clean up and leave at the end, without delaying service to the following customer. Smooth, efficient, and more pleasurable for everyone. But yes, it takes up more room in the store and may require restructuring the original equipment.

The design principle here is recognizing that customers need space and time to get ready for a transaction, then more space and time to clean up after finishing. By providing two spaces, two buffers, one customer does not delay the next.

Temporal Double Buffering: Checkout Lanes

The two-sided checkout lane is a spatial double buffering, with one buffer on each side of the cashier. A second use of double buffering is temporal, providing sufficient linear space to segregate the operations: setup, tallying the items, and cleanup, thereby allowing the next customer to begin loading the setup buffer even before the first customer has finished. A good example of this is supermarket checkout lanes.

Supermarkets often use linear space to separate setup, tallying, and cleanup. An automated belt moves grocery items from the setup location to the tallying location. The belt is long enough to hold the items for several customers, often with a separator between them. When one customer's things are tallied, the belt moves the next customer's things up to the cashier, freeing up space for the following customer to unload. Moreover, as each item is tallied, it is then moved to a large cleanup area where either a second employee or the customer can pack the purchased goods, allowing the cashier to deal with the next customer.

Temporal Double Buffering: Drive-through Restaurants

Drive-through restaurants use a temporal double-buffering scheme. Customers drive their cars up to an ordering window and place their order. Then, they drive forward to the takeout window, which can be deliberately made long, sometimes even requiring the driver to go around the corner of the building. This serves two purposes. First, the car clears the ordering window for the next customer. Second, the time required to drive between windows gives the staff sufficient time to prepare the order. This linear separation into two steps of ordering and then receiving and paying also allows room for two queues: one to wait until the order is placed (where the waiting time can be useful, giving customers time to peruse the menu and make their decisions), the other queue to wait for the order fulfillment and payment. Some processes are made more efficient by inserting a separate payment location prior to picking up the completed order.

Temporal Double Buffering: Coffee Shops

Many coffee shops and fast food locations use a form of linear, temporal double buffering by having orders placed at one window and picked up at another. Once again, this separates the operations, allowing for greater efficiency. People placing orders are not delayed by the wait for previous customers to pay for and pick up their food. In addition, the separation gives space for several queues, especially important because in these places, the orders might not be filled in the same order as they were received. Items that are quickly prepared can jump ahead in the pickup queue. Complex items can be delayed. A linear, single line would have to move at the rate of the slowest item.

Designing the Lines

One Line Feeding Multiple Servers

Suppose there were ten servers (clerks, cashiers, or ticket takers) serving a crowd. If the crowd were divided into ten lines, each line would be one-tenth the size of a single line, but it would also move one-tenth as fast. But if there were only one line, when people reached the head they could go to whichever of the ten cashiers were free. In this case, the one line would move ten times as fast as a line with only one server. One line with multiple servers provides the fastest-moving lines, as well as the most fairly perceived situation. Make that one line have multiple turns and the result is ideal: a fast-moving line that is visually short.

This system has its own set of issues to be solved, with the details varying from situation to situation. This analysis could go into great depth: an entire book could be written on the various ways of managing lines of customers. With proper observation of the bottlenecks and issues to be overcome, the service can often be made both more efficient and more pleasurable. Efficiency need not come at the expense of overburdening the clerks or underserving the customers.

Customers far prefer the perceived fairness of a single line feeding multiple servers rather than individual lines in front of each server. As mentioned previously, a single line moves more quickly than multiple lines, even though the total numbers of customers served at any time is the same. But the perceived fairness of the single line is dramatically greater.

The major difficulty with the single-line scheme is directing people to the correct server. If there are many servers, it isn't always easy to tell when one is free. Sometimes this is done by the people in line themselves, with those near the front of the line always eager to tell others when a server is free. Even when the previous customer has left a server, that position may not yet be available, so it is necessary to wait for some signal. Inefficiency increases as one by one customers wait for a clear signal to proceed, then walk over to the available server, unload their materials, and initiate the transaction. This calls for yet another double-buffer solution.

In some situations, the double buffering is aided by having an employee act as a line manager, directing each person to the next available line. In some cases, I have seen this done so as to form a deliberate second queue in front of each clerk, usually just one or two people long. This is often found in airport immigration and customs areas. By always having one or two people in line in front of each clerk, the startup time for the next person is minimized, although there is some risk that the one person in front will be the one with a complex, time-consuming transaction, leaving the one person left in the queue feeling unfairly served. An alert line manager can solve this problem by moving that person quickly to another line.

All sorts of clever variations on this procedure have been introduced, including the use of electronic signaling systems to let customers know which clerks are available. I've seen this done with flashing lights and with display screens that have arrows pointing in the correct direction, giving the name or number of the target location.

Number Assignment

Providing numbers to arriving customers, sometimes differentiated by the type of service required, is a version of the single-line, multiple-server solution, but in this case the customers can sit or walk around rather than stand in line. It can be found in busy places like banks and government offices. This system also has the advantage of allowing different classes of customers to be served differently. Departments of Motor Vehicles often use this scheme. As people enter the building, they encounter a clerk who determines their needs, assigns them to the appropriate queue, and gives them a number that specifies their place in that queue.

People waiting for a driving test are in a different numerical queue than people who simply need a form, who might be in a different queue from people waiting to renew a license or to submit a form. The numbers themselves provide feedback, so people can track both the rate of progress of the queue and also how far away their number is from the number currently being processed.

Of course, electronic variants are possible, which includes handing out pagers to people so they will be buzzed when their time comes. The electronic variants have the virtue of giving people more freedom to wander, but they eliminate the feedback that comes from being able to observe the length of a line or the current number being served.

Targeted Admission Times

One way to minimize the trauma of waiting lines is through reservations. But this has to be done in a way that seems fair and reasonable, even to those without reservations. That is, people have to believe that they too could have enjoyed the benefit of a reservation had they planned ahead. A modification of a reservation system is to provide each person with an admissions ticket with a guaranteed time, even if it is for some time in the future. Then, instead of waiting in line, they can do other activities and not show up until the system is ready for them.

This is the philosophy behind restaurant reservations. It is also how entertainment parks sometimes handle long lines for rides: as people sign up, they are given an electronic device that lets them roam freely and do other activities, but that will call them back in time for them to get to the ride just as it is ready to receive them. Restaurants also often do this with waiting patrons: sign up at the counter and get a paging device. When your table is ready, you are paged, and the resulting buzzes, light flashes, and vibrations notify you that it is time to gather your party and go get seated. These systems all have their own set of problems, but all of them are design efforts to get around the issue of long, uncomfortable lines.

Disney theme parks provide a special pass, the Fastpass, to avoid long lines. Everyone is entitled to this, but only one pass can be held at a time. It doesn't let people go ahead of the line: it is a guaranteed service time. Here is how it works. When people get to the ride, a sign tells them at what time the Fastpass slot will become available. People holding a Fastpass can do anything they wish, as long as they come back to the ride within an hour after the printed time. They still have to wait, but most of that time is spent wandering the park, perhaps even taking other rides (for which they will have to wait in line because of the one-pass restriction).

When the people return to their Fastpass ride and get into the special line, it is short and fast. Other people waiting in the longer, regular line, do not feel cheated: they know that they too had a choice of getting a Fastpass but chose not to. It is very important for this perception of fairness that only one Fastpass is allowed at a time. This is enforced rather simply: the park admissions ticket is inserted into the Fastpass machine, which identifies the customer and checks eligibility before issuing each Fastpass.

In a neighboring theme park, Universal Studios Orlando, people can choose to purchase an expensive Front of Line Pass that can be used for any ride at any time. This pass causes offense. One person whose family had visited both Disney and Universal on the same trip explained that the Disney system seemed fair and equitable, whereas he and his family were quite annoyed at Universal: "the rich get to go first," he said, "and that isn't fair." "Harpo," the screen name of someone writing on the Atari community forum, called it "obnoxious. I resent that only those who are willing to shell out more money after already paying for admission are able to use them."

Memory Is More Important than Reality

Which is more important: the experience during an event or the later recollection of that experience? In the abstract, the question would seem difficult to answer, but consider that your future behavior will be controlled by your memories. Memory is by far the more important aspect of the waiting line experience, one reason the ending experience is so much more important than the beginning or middle. The memory of an event can be more important than the actual event.

Research on human memory demonstrates that recollections of events are active reconstructions of the experience, subject to many possible distortions. In the legal profession, the unreliability of eyewitness testimony is well known and many psychological experiments have shown how easy it is to distort one's memory of events. Consider the woman who remembered with great pleasure her visit to Disney World in Orlando, Florida, recalling the wonderful Disney characters she interacted with: Bugs Bunny, Cinderella, Mickey Mouse. Except that Bugs Bunny is not a Disney character, and therefore could not have been part of her experience. She had taken part in a psychology experiment where she was asked to recall her actual visit to Disney World right after she had been shown an advertisement for Disney World that showed Bugs Bunny.

Moreover, the memory of the whole experience is more important than the experiences of the separate parts. Researchers Richard Chase and Sriram Dasu at the University of Southern California's Marshall School of Business point out effective strategies for situations with mixed positive and negative components, including finishing strong, segmenting the pleasure while combining the pain, getting bad experiences out of the way early, and building commitment. The results of these and numerous other studies of people's memories of events reinforce the basic design principles: manage the ending, provide mementos to take home, start strong and end strong, and bury unavoidable unpleasant aspects in the middle.

Bob Sutton, Professor of Management Science and Engineering at Stanford University, has suggested that an important component of participants' memories of an event comes through the photographs they took. Thus, providing photographic moments in the waiting line—for example, the family happily engaged with one of the park's costumed characters—ensures that the record the family brings home with them contains the positive moments of their visit. With each viewing of the pictures, the family enhances their positive memories without reawakening the negative ones.

Although waiting is universally disliked, there are times when it is useful to induce waits artificially. Traffic lights are a good example of introducing a deliberate wait for one set of vehicles, the better to permit other vehicles or people to gain access.

In theme parks the waits are deliberate. "What else would we do with the people?" I was once told by a high-level executive of one of the major theme park companies. "It is too expensive to add more rides." Waits are unavoidable when there are more people than resources, so in this case, although the waiting was deliberate, the company's response was to make those waits as enjoyable as possible.

Waits can be used to enhance pleasure. We wait until dinnertime to eat, in part for cultural reasons but also so that we are hungry again. We refrain from opening gifts before the allotted time, the wait increasing our suspense. We sometimes welcome waits, for they allow us time to savor the moment, or to read, finish a conversation, or complete a desired activity. Some waits at the start of an activity are beneficial, allowing us time to prepare. At restaurants and even fast food places, the wait gives us time to study the menu and decide on our choices.

There are even times when waits are too short, when we are forced to respond before we are ready, or when we did not have enough time to finish the intervening activity.

As I have already discussed, the pleasantness of the wait can be manipulated by adding distractors—tasks to occupy people. Waiting rooms provide magazines and television. Some banks have installed television screens for those waiting in line. It is rumored that waiting for elevators has been made more pleasant by adding full-length mirrors near-by, so people can examine themselves while waiting. Airports have expanded their waiting areas into full-fledged activity centers, with shopping malls, televisions, restaurants, and bars. At certain international airports well known for their extensive collection of excellent stores, some people even plan their trips to extend the time between their flights.

Note that there is a paradox built into the way people perceive time: unfilled time is perceived as lasting longer than filled time at the time of the experience, but when later recalled, unfilled time is perceived as being shorter than filled time. So which should be given to customers?

The way to answer the question is to realize that what truly matters is the total experience. Although short waiting times are to be preferred to long ones, if the time is filled with interesting activities, then at the time it is experienced, it is perceived as going quickly and enjoyable. Later, when the activity is being remembered, the events will dominate, and as long as those events are pleasurable, the end result can be positive: "Yes, we had to wait for a long time in line but the wait was fun."

Any place that has numerous people waiting for extended periods could follow this practice. But note that this only works if one's place in line is assured. Here is a case where an attempt to make an experience more pleasant can have the opposite effect if it is accompanied by worry about missing an event or losing one's place in line. Some complexity has to be added to simplify the experience: number assignment, reservations, or targeted admission times will help. Even so, people have been known to miss their flight because they were so distracted by the available activities at the airport.

Waiting in line is never an end in itself. It is always done in order to gain access to something else. The memory of the line can be enhanced both by adding positive experiences during the wait that can later be favorably remembered, but also by making the events at the line's termination be incredibly positive and well worth the effort. In fact, through the psychological mechanism known as "cognitive dissonance," the suffering actually enhances the enjoyment of the later event. Although the dissonance reduction is subconscious, think of it as the subconscious mind deciding that "any event that requires so much effort to enter must really be important and wonderful." Cognitive dissonance was first proposed in the middle of the twentieth century by Leon Festinger to explain how people manage when events contradict a basic strongly held belief. To Festinger's initial surprise, such contradictions often seemed to intensify the belief, rather than demolish it. The theory of cognitive dissonance explains why this might happen.

Disney theme parks are probably the champions at handling the dislike of lines. When I query people about their trip to a Disney theme park, I ask two questions: What did you dislike most? Would you go again? The answer to the first question from people in the United States, Asia, and Europe, is immediate: the lines, the queues, the waiting—the description varies with the location in the world, but the meaning is always the same and always immediate, without any need for thought. People dislike the lines. But the answer to the second question is much more revealing. "Would you go again?" "Yes!" comes back the answer, again immediately, without any need for thought. People may dislike the lines, but Disney manages them so that they seem appropriate, fair, and necessary.

When Waiting Is Handled Properly

I often ask people about their experiences, which include numerous situations where people have to wait in line. Waiting for a train, waiting for a restaurant table, waiting in line to be served at a university cafeteria: all these waits are accepted as reasonable and fair. Waits that are deemed unreasonable are often those where fairness was violated or the rules of behavior not well stated.

Thus, at a movie theater complex, with many ticket windows but no clear lines, people were uncertain how to behave and, as a result, not happy with the experience. The same comments were heard from a lineup of people waiting to be served at a marketplace. It was unclear how to behave and it always felt that people who arrived later were being served earlier. The uncertainty produces anxiety and other negative emotions. Those who know how to maneuver in such situations may feel a certain amount of pride in their ability to get served, but these positive feelings on the part of some are often outweighed by the negative feelings on the part of others. Note too that you cannot assess the strength of the negative feelings just by asking those who are waiting to be served. The people with the strongest negative reactions may have stopped attending altogether.

While writing this book I was an unwilling participant in a situation that illustrated the importance of communication. I boarded an airplane that was scheduled to take me from Chicago to San Francisco, but the departure was delayed while airline technicians paraded back and forth down the aisle to the rear of the airplane. Frequent announcements told us that the rear toilets were not working, and that as soon as they were fixed we would leave. Then we were told that we might leave without them operating. Every twenty minutes I received a text message telling me of the revised departure time. After an hour of continual maintenance efforts and announcements, a pilot explained that he had decided that we should not fly with only one operating toilet, but instead we would disembark and another airplane would be found. Despite the uncertainty, the passengers were calm and understanding. My seatmate told me that it was reassuring that the captain himself had made the final announcement, explaining his reasoning. The continual communication was helping everyone feel informed and confident that they were in competent hands.

Once we got into the boarding area, though, the situation changed dramatically. The gate agents were bombarded with questions by travelers, but they had no information. An agent announced a gate change using the correct flight number but the incorrect destination. I quietly corrected her and she explained to me that she had been called in hastily to help and wasn't clear what the issue was. As we milled about aimlessly, passengers fretting over missed appointments and missed airline connections, it was clear that the gate agents were even more stressed than the passengers. At one point an agent tried to make an explanatory announcement, but her confused statements puzzled the passengers so much that they interrupted to ask questions. From my vantage point, they were asking sensible questions in reasonable voices, but the flustered agent said that if they didn't stop she would call the police. When one more question was asked she indeed did pick up the telephone, but evidently had second thoughts and simply left the scene. "She's ready to explode," the person next to me said. "I'm glad I don't have her job."

Why the difference in behavior? Lack of information and of appropriate feedback, and lack of understanding of the underlying causes. The passengers, and for that matter, the employees, lacked a clear conceptual model. Note that the stress appeared to affect the employees more than the customers. Their situation was worse because they had to endure the complaints, even though they had neither any part in causing the situation nor any knowledge about the solution. Informed, intelligent feedback is as important to the staff as it is to the customers.

Designing the Experience

Emotions color our experiences and, more importantly, our memories of our experiences. Emotion affects people's judgments. In *Emotional Design* I summarized much of the research with the statement: "Attractive things work better." Wash and polish your car and it drives better; shower and dress up in fancy clothes and the world looks brighter. Obviously, washing a car does not make it any better mechanically, but it makes one's perceptions change. The same lesson applies to the way we deal with complex things. When we're in a positive mood, minor difficulties or confusions are considered minor, not a major problem. But when we're anxious or irritable, the same minor setback can become a major event.

Susan Spraragen, a research scientist at IBM, has been studying the emotional states introduced by service experiences: we first encountered her work in chapter 6 (see the expressive service blueprint of figure 6.3). Spraragen provided me with the illustration used in figure 7.2 to show the frustration that can be felt when the wait seems inappropriate. The person in figure 7.2, a patient, feels ill ("lousy") and wants to speak with a physician or nurse. But instead of getting help, patients must first identify themselves, wait to be found in the medical clinic's database, and then confirm the status of insurance coverage. "Is anyone listening?" thinks the patient as the clinic employee tries to access the medical records. To the patient, the questions by the clinic staff add complications to a simple call for help, and the resulting delays lead to frustration and irritation. This heightened emotional state does not help either patient or staff.

Although there are many legitimate reasons why the clinic might first need to find out who the patient is, access records, and check on the patient's health insurance, to the patient, it all appears to be an unnecessary barrier. The feeling could very well be shared by the person who is interacting with the patient and who would prefer to be administering health care more directly. The situation is especially difficult in medical situations where people are likely to be in emotional distress even before they encounter the medical staff. In the case illustrated in figure 7.2, where the patient starts off "feeling lousy," this state has probably sensitized the emotional system to be more upset than usual to delays and difficulties. Special design is necessary in this case: perhaps the medically relevant questions could be asked first and the identifying information requested only as the appointment is being scheduled.

The impact of emotion has numerous design implications. Make the surrounds bright and cheery, attractive and inviting.

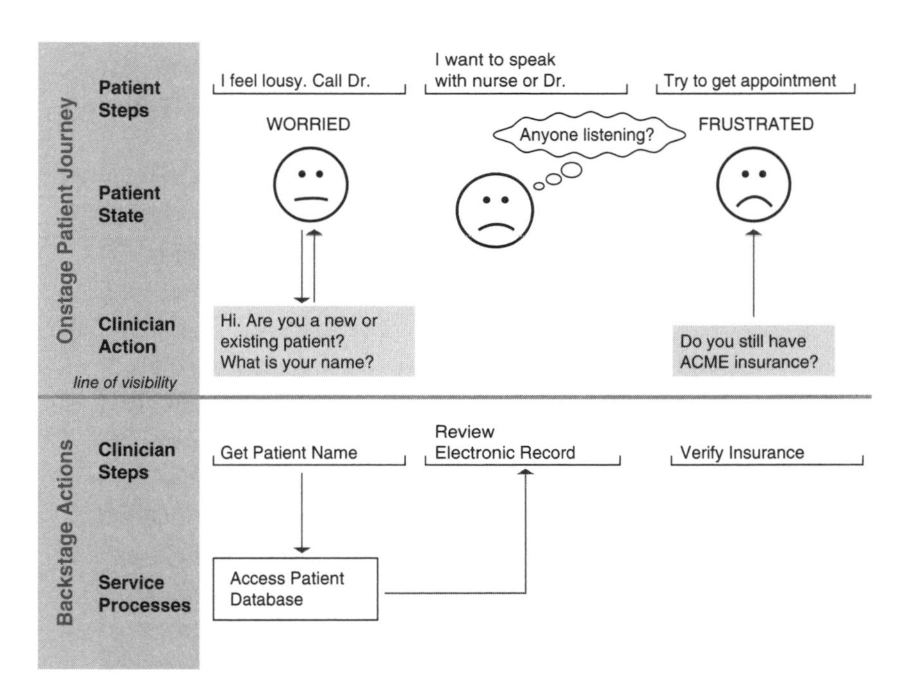

Figure 7.2
A simplified expressive service blueprint. This service blueprint illustrates the frustration and anger that can arise while waiting, especially when the reason for the wait is not explained and, worse, seems like a diversion from the problem of "feeling lousy." From Spraragen 2010, used with permission.

Make sure that everyone is in a positive, helping mood. The environment is not just the physical surroundings; it includes the employees and other customers. Employees have to be seen as cheerful and helpful, and teaching employees how to be that way, especially after a long shift of high-stress interaction with numerous unruly and upset customers, families, and children, is a worthy design challenge in its own right. Even so, the demeanor of the employees can make a dramatic difference on the customers' impressions. Similarly, it is important to overcome the negative emotions of upset customers. I have been told that "Disney employees are taught to pay special attention to customers who are most upset, both because they are unhappy, and especially, because negative emotions can spread, an observation very consistent with the vast research on emotional contagion."

These same lessons apply to the medical clinic, even if more difficult to handle because of the underlying level of stress caused by medical problems, the uncertainty faced by all involved, medical staff and patients, and the crisis mentality that necessarily surrounds some wards, most especially emergency rooms. Nonetheless, the situations can be improved. The environments can be made more attractive, with more attention paid to the waiting experience, and the procedures should be designed to be understandable and appear appropriate. Special care should be expended for the well being of the patient and any accompanying entourage of friends and family. Even though a number of nonmedical administrative routines are required, these should be secondary to the consideration of the emotional state of the patients and, for that matter the medical staff.

Emotions are contagious. When people are happy and smiling, others around them will be happy and smiling. When people are nervous and irritable, others around them will follow. Get people in a good mood and keep them there. Emotions dominate everything else.

Waiting is a simple activity that complicates our lives. But there are ways of diminishing the frustration and boredom, ways of helping people pass the time. The six principles for the design of waits provide suggestions. For example, suppose, while waiting for their baggage after a flight has arrived, passengers could view television monitors that displayed the progress of the baggage from the airplane cargo hold to the waiting carts and transport to the terminal and final placement on the conveyor belts. The backstage operations of many companies can be interesting to customers. Why not let the waiting people see what is happening? Coffeehouses do this by letting customers watch the baristas. Sandwich makers do this by letting customers watch and direct the person making their sandwich. This principle even works without a physical presence. The Web site of Domino's Pizza lets people trace the progress of their order, including the names of the cook and the delivery person, along with the expected time of arrival. The conceptual model is clear and direct, and the feedback transforms what could be an annoying wait into a personalized adventure.

After an event, all one has left are memories of it. Because most waits are en route to a desired outcome, it is the memory of the outcome that dominates, not the intermediate components. If the overall outcome is pleasurable enough, any unpleasantness suffered along the way is minimized. Terence Mitchell of the

University of Washington's Foster School of Business and Leigh Thompson at Northwestern's Kellogg School of Management call this "rosy retrospection." Mitchell and colleagues studied participants in a twelve-day tour of Europe, students going home for Thanksgiving vacation, and a three-week bicycle tour across California. In all of these cases, the results were similar. Before an event, people looked forward with positive anticipation. Afterward, they remembered fondly. During? Well, reality seldom lives up to expectations, so plenty of things go wrong. As memory takes over, however, the unpleasantness fades and the good parts remain, perhaps intensify, and even get amplified beyond reality. The memory of an event is far more important than the reality. It's all a matter of design.

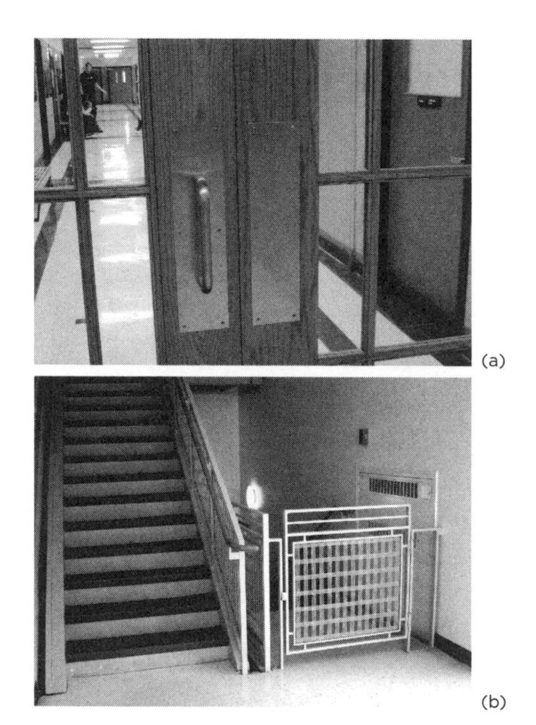

Figure 8.1
Signifiers that are effective communicators. In (a), the grabbable handle provides a clear signal that it is to be grasped by the hand and pulled whereas the flat plate offers only one affordance—to be pushed. The plate is an effective signifier both of the need to be pushed and where it is to be pushed. In (b), the open staircase invites usage: the architectural design serves as an effective signifier. But the gate blocking the stairs that go down signifies restricted access. It is still possible to open the gate and proceed, but the purpose of the gate is to act as a forcing function, making it less likely that people fleeing the building in an emergency will run down the stairs into the basement rather than exit when they reach this level.

8
Managing Complexity
A Partnership

How to Start a Ford Model T Automobile

"Now you ready? Spark retarded, gas advanced. Spark up, gas down. Now switch to battery—left, remember—left." . . . "Hear that? That's the contact in one of the coil boxes. If you don't get that, you got to adjust the points or maybe file them." . . . "Now this-here is the crank and—see this little wire sticking out of the radiator?—that's the choke. Now watch careful while I show you. You grab the crank like this and push till she catches. See how my thumb is turned down? If I grabbed her the other way with my thumb around her, and she was to kick, why, she'd knock my thumb off. Got it?" . . . "Now," he said, "look careful. I push in and bring her up until I got compression, and then, why, I pull out this wire and I bring her around careful to suck gas in. Hear that sucking sound? That's choke. But don't pull her too much or you'll flood her. Now, I let go the wire and I give her a hell of a spin, and as soon as she catches I run around and advance the spark and retard the gas and I reach over and throw the switch quick over to magneto—see where it says Mag?—and there you are."
—From John Steinbeck's *East of Eden*

Complexity is both necessary and manageable: that is the message of this book. But complexity can overwhelm and frustrate us. So what should designers be doing to tame the complexity? And how do we cope with the complexity that remains? We've already covered the fundamentals: now is the time to put them together. It is important to recognize that this is a partnership between the designers and us. Designers can do their part, organizing and structuring the systems we deal with so that we can understand and learn them. But we too have to do our part: simplicity, after all, is in the mind. Complex things become simple after we have mastered them, after we understand how they operate and the rules for interaction. After the designers have done their part, we need to do ours: to take the time to learn, understand, and practice. Through this partnership, complexity can be managed.

Automobiles once were incredibly complicated, as the opening quotation for this chapter illustrates. The author John Steinbeck commented on his own description of the difficulties:

> It is hard now to imagine the difficulty of learning to start, drive, and maintain an automobile. Not only was the whole process complicated, but one had to start from scratch. Today's children breathe in the theory, habits, and idiosyncrasies of the internal combustion engine in their cradles, but then you started with the blank belief that it would not run at all, and sometimes you were right. Also, to start the engine of a modern car you do just two things, turn a key and touch the starter. Everything else is automatic. The process used to be more complicated. It required not only a good memory, a strong man, an angelic temper, and a blind hope, but also a certain amount of practice of magic, so that a

man about to turn the crank of a Model T might be seen to spit on the ground and whisper a spell. (Steinbeck 1952)

The automobile stands as an excellent example of the partnership between those who design and those who must cope. Designers and engineers have dramatically simplified the operation of the car, but drivers have to do their part as well. Most people have to take driving lessons that combine classroom lectures and driving practice along with an official examination. Even after passing the examination, the novice driver may take months or years to become skilled at the task.

Designers can transform otherwise confusing systems into understandable ones. But if the systems are dealing with complex activities, that doesn't mean that the result will be immediately understandable and usable. In the end, the burden is on those who use them. Even simple tools take time to master: the lowly screwdriver, wrench, hammer, potato peeler, or the pencil are all as simple as can be imagined, but all take practice before they are mastered. Taming complexity is a partnership between those who design and those who use.

Computers are often blamed for the complicated nature of modern life, a complaint that has merit. But computers also offer the potential to simplify life. The modern automobile is a good example of appropriate design, where hundreds of computer chips, sensors, and motors work in the background to regulate the fuel–air mixture, prevent skidding, maintain stability, and warn of potential dangers. These embedded computers do not require conscious attention or control: they monitor the automobile, the driver's actions, and the environment, and respond accordingly.

Modern cars can even communicate with one another. They can monitor traffic and weather, suggesting traffic routes that take account of speed limits, construction, and traffic conditions. The automobile and its computer systems become more complex every year, but this hidden complexity transforms the driver's task, simplifying it while making it safer. This is yet another application of Tesler's law of the conservation of complexity, discussed in chapter 2.

The Fundamental Principles for Managing Complexity

We need two sets of principles for managing complexity: one for design, one for coping. In the end, the rules all evolve around communication and feedback. The design must include appropriate structures to aid human comprehension and memory as well as tools for learning, and for handling unexpected events. The task is made more difficult by factors outside the control of the designer. The system might be used alongside other systems that do similar things but follow different design principles, and even though each alone might be sensible and understandable, the contradictions complicate the life of the person who must use both. Moreover, the design has to support usage in the face of the unavoidable interruptions of life.

Rules for the Designer: Taming Complexity

The preceding chapters provide a number of rules for designers. The fundamental requirement is to make things understandable.

Good conceptual models are essential, but they aren't useful unless properly communicated. The design tools are conceptual models, signifiers, organizational structure, automation, and modularization. In addition, the design team needs to provide learning tools: manuals and help systems.

The major path to good, usable design is communication. Once upon a time the word "design" referred primarily to appearances: automobile styling, fashions, and interiors. Products were viewed in photographs, prizes awarded solely on the basis of appearance. Today that has changed: the design world now is concerned with function and operation, with fulfilling fundamental needs, with delivering positive, enjoyable experiences. We now recognize that one critical component of good design is good interaction, and interaction, to a large extent, is about proper communication.

When the field of human-centered design was just emerging, two Swiss researchers, Jurg Nievergelt and J. Weydert, argued for the importance of three knowledge states: sites, modes, and trails. Their insights can be translated into three basic needs: knowledge of the past, the present, and the future.

Knowledge of the present means knowing the current state: what is happening right now? Where are we with respect to our goals and starting point? What actions are now possible? It is amazing how many systems do not give a clear indication of their current situation.

Knowledge of the past means knowing how we got to the current state. Some systems erase the past, so if we find ourselves in an unexpected or undesirable state, we may not know how we got there. We may not even remember what the previous state was. As a result, if we like the current state, we may not be able to remember how to get back in the future. If we do not like the current state, we may not know how to reverse the action so we can return to the previous state.

Knowledge of the future means knowing what to expect. Our actions are based on expectations of the future. Many of our emotional states are driven by expectations. The lack of knowledge of future expectations not only makes the task difficult but it leads to unnecessary tension.

One of my fundamental design rules is to avoid error messages. After all, the natural world manages quite well without error messages. To me, good design means never having to say "that was wrong." An error message really indicates that the system itself is confused: it doesn't know how to proceed. It is the system that needs to be scolded, not the person.

Life has no error messages. Similarly, computer and video games are examples of complex systems that work well without error messages. When a person tries something not understandable by the system, it simply does not respond. It's like pushing a door to open it when it really should be pulled. There is no error message. There is no scolding. It simply doesn't open. In the case of a door, it isn't too difficult to figure out how to proceed. With many of our complex systems, their workings are not visible, so when we try something and nothing happens, we may not have any idea of how to continue: here is where some help is useful—but it should be thought of as assistance, not as errors. But far better than forcing a person to ask for help would be systems that were self-explaining. We usually can figure out what to do with physical systems because everything is visible so the alternatives are clear. Electronic, computer-based systems need to do the same, to present sufficient information about how they work so that when something goes wrong, there is perceptible evidence of the problem and the possible alternative actions. Once again, information about the past, the present, and the future.

Errors are wonderful teaching opportunities. When the system cannot proceed because the person has provided ambiguous, erroneous, or incomplete information, instead of flagging it as wrong, present an explanation of the problem along with the necessary tools to fix it, right there, on the spot. When a person understands, not only can the problem be avoided, but the system becomes conceptually simpler.

All modern theories of learning emphasize the importance of active construction and discovery on the part of the learner that is enhanced by coaching, mentoring, and guided learning. The best time to learn something is just after the person has discovered it is needed. That's when the demonstrations, tutorials, and explanations are most valuable. Try to teach too soon and the result is boredom and disinterest. But catch people just as they need the material, and they are apt to be highly motivated, attentive learners.

"Divide and conquer" is an old strategy with powerful implications for design. When there are many pieces, the structure can be made modular so that only the pieces of concern at the moment are in focal attention. Grouping and organizing provide a structure for understanding complexity.

All of these rules are really based on two ideas: meaningful communication and a compelling conceptual model.

Signifiers

As we have seen throughout this book, a signifier is any perceivable sign for appropriate behavior, whether intentional or nonintentional.

It is a powerful tool for designers to enable communication in a natural, comfortable fashion, comfortable both for the designer and the person using the design. Signifiers act like a natural part of the world, and so the communication can be effortless and appropriate. People use the world as a vast database of information, guiding them through the activities of the day. Much of the information needed for guidance is there, sometimes as explicit physical information, sometimes implicit, sometimes as social indicators of appropriate behavior. Four simple but effective signifiers can be seen in figure 8.1.

Signifiers are powerful design tools. Designers already make use of them. Unfortunately, the signifier is often confused with the closely related concept of "affordance." An affordance is a relationship: it speaks to the possible actions a person can perform upon an object. The concept was first developed by the perceptual psychologist J. J. Gibson, who applied it to all organisms and all environments. Affordances, to Gibson, were relationships between potential organisms and potential objects that exist in the real world whether or not anyone was aware of their presence.

In 1988, I appropriated the concept of affordance for the world of design. Although the concept was readily accepted and is now widely used, it is frequently misunderstood. To Gibson, an affordance exists whether or not anyone ever notices it. To the designer, if affordances are not known, then they might as well not exist. In other words, the designer is primarily concerned with *perceived* affordances; the perception is critical. As a result, when designers correctly observed that some people were having difficulty using a product because they failed to notice the affordance, they would add visible signs of its existence. But, lacking the appropriate vocabulary to describe what they had done, they would say that they had "put an affordance on a product" when in fact they were making visible the presence of an already existing affordance. What they really were doing was adding a signifier. The designers had no choice; no other word existed to describe what they had done (the word "signifier" had not yet been introduced), so with time, the term "affordance" in design has come to mean something perceptible.

I strongly urge the design community to distinguish between affordances and signifiers. In most cases, the word affordance should go away, for invariably the designer cares only about what can be perceived, which means the signifiers. Note that all perceived affordances and signifiers are methods for communication. The art and science of selecting appropriate signifiers is an important design skill: good designs have signifiers that are perceivable and informative as well as being aesthetically pleasing and harmonious with the rest of the product.

Want to find bad design? Want to find a lack of appropriate signifiers? Look for signs explaining how to use something. For example, "push" or "pull" labels on doors that, if properly designed, would not need them, and posted notes and handwritten signs, instructing people how to use devices that are poorly designed. All these markers and add-ons are actually social signifiers, aids added by one group of people to benefit another.

Affordances are important, for they are the part of the world that makes actions possible. Designers are still responsible for ensuring that the objects and systems they design have the proper affordances, but if they are not noticed or perceived, then they will probably fail to achieve their purpose. The designer must communicate the range of actions through signifiers. Signifiers are critical to effective communication.

Structure

One way to simplify an otherwise complicated situation is by adding structure. Structure the task into manageable modules, where each module is simple and easily learned. This is the secret of the silversmith's planishing hammer discussed in chapter 2. The silversmith's workbench (figure 2.4c, p. 44) looks complicated, but it was acquired tool by tool, where learning any individual tool is a manageable and understandable task. The result is that to the silversmith, the complex-appearing bench is viewed as an understandable, sensible collection of numerous simple tools.

Another way to simplify is to reconceptualize. To reconceptualize is to find a different way of framing the problem. A good example comes from the major technological change in the way that we record television shows.

The original technology that enabled people to record television shows as they were broadcast was the "video cassette recorder," called a VCR. These VCRs were so poorly designed that many people could not even figure out how to change the time on the clock. The difficulty of setting the correct time became such a national joke that in 1990 President George H. W. Bush announced at a press dinner in Washington: "We have a vision; by the time I leave office I want every single American to be able to set the clock on his VCR." (He failed.)

Recording a show was even more daunting. For example, to record a show that was scheduled to be broadcast on channel

37 on Wednesdays, from 9 PM to 10 PM, you would first have to make sure the VCR clock was correctly set, then go into programming mode and tell the VCR to set itself to channel 37 on every Wednesday at 9 PM, and then to start recording for precisely sixty minutes. Of course, you first had to look up the broadcast schedules in a newspaper or television guide.

The secret to success in taming the complexity of the VCR was not through clever, sophisticated user interface design, but rather in recognizing that the problem had been approached from the wrong point of view. People wanted to record a show to watch at another time; they had no interest in when the show was actually broadcast. Why should they have to set the day or the time or the channel?

Today, most video systems have completely reconceptualized the task. Now people can indeed record the shows they want simply by entering the shows' names and the system does the rest. Now people program their video systems without ever being aware that this is what they are doing. In many cases, no recording is even necessary: shows are available whenever the viewer wishes to see them, much as books in the library or websites are always available, ready for viewing whenever the interest strikes. In many cases, the best way to simplify a task is to reconceptualize it.

Modularization: Divide and Conquer

One form of structure is modularization: dividing a complex structure into a number of small, manageable modules. That's how well-designed multifunction printers, scanners, copiers, and fax machines do it: each function is compartmentalized by grouping and graphics, so each is relatively simple.

One of the complexities of life is the control of our entertainment systems. In part, this complexity is required: the modern system serves many functions, including the viewing of photographs and home videos, Internet sites and videos, retrieving and playing material from photo, music, and video libraries, playing videos and games, serving as picture frames, displaying favorite family scenes when the displays are not otherwise being used, and even for watching television and listening to music or the radio. As a result, there is a multitude of different devices that must all be connected to one another, each of which must be controlled. The result is a confusing medley of remote controls, any single one being barely comprehensible and the total package insufferable: see figure 8.2a.

Figure 8.2 shows how good design can make an otherwise complex system appear simple. The mistake that designers of entertainment systems make is in believing that the people using the system wish to control the individual components. As a result, we are given complex control devices, each capable of many functions, but with only slight attempts to provide a clear, comprehensive conceptual model of how these are to be used in actual operation. Many of the controllers recognize that the person will have multiple devices, so their designers attempt to provide a "universal" control, one product that can control multiple devices. However, because they still focus on the devices themselves, the result is to increase both the perceived and the actual complexity. Logitech's Harmony remote control overcomes these problems.

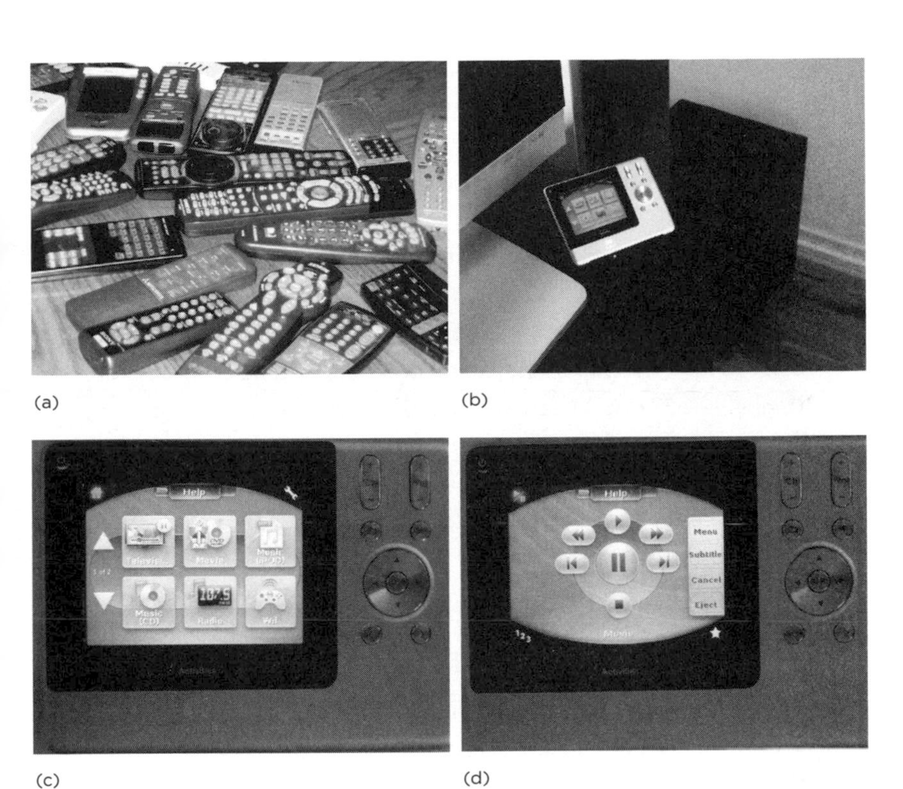

Figure 8.2
Simplifying through good design: (a) shows the remote controls I own for all the devices in my entertainment center that need controlling: this daunting set of complex controllers is unusable. Complex, unintelligible devices for a complex task. (b), (c), and (d) show Logitech's Harmony remote control that I now use to control my system. It overcomes the complexity while offering the same capability. Good design makes complex things simple to use.

The controller shown in figure 8.2b, c, and d takes an activity-centered approach. That is, the operation is centered not on control of the DVD player, or the radio, or the game machine, but on the activity: watch a movie, watch TV, or listen to music. To use the system, an activity is selected (figure 8.2c), and then the control screen changes to reflect the requirements of that activity (figure 8.2d shows the screen after the activity of "Movie" has been selected). The mechanical controls on the right are for items required for most activities: the control pad, volume control, and channel selector (not used during movie watching), as well as the mute button. Activity-centered design models the actual requirements of the viewers, thus properly modularizing the otherwise complex set of controls, reducing a complex pile of remote controls to one elegant, simple one. It is all in the conceptual model, the appropriate modeling of the task.

David Kirsh, in the cognitive science department at the University of California, San Diego, has examined how people structure their environments to simplify their tasks, organize their actions, and remember and organize their places after interruptions. He calls this work "cognitive congeniality."

Kirsh has shown how intelligent placement of objects can put much of the burden of remembering on the world, simplifying the cognitive load that falls on the person. Imagine preparing a salad for dinner, with numerous vegetables to be washed, peeled, and cut. Experienced cooks separate the washed and unwashed vegetables by placing them in separate areas. At any moment a quick glance shows how much more needs to be done. If the cook leaves the kitchen, it is easy to remember where to continue working upon returning. On the face of it, this placement of items

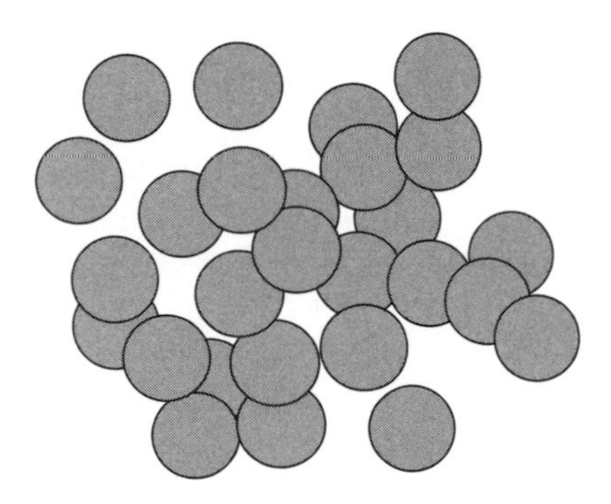

Figure 8.3
Count the objects. Don't point, don't use any aid—just look and count. This is a very difficult task—it is not cognitively congenial. Drawn following the description of the task in Kirsh 1995.

seems trivial, but the underlying philosophy is powerful. The artist's or jeweler's bench will show similar patterns.

Some arrangements highlight the obvious actions to be done, some bring attention to opportunistic acts, and some deliberately hide or make obscure actions that are undesirable (e.g., moving the filled container off to the side to avoid accidental spilling). Space is used to cue the ordering of operations, as reminders, to occlude undesirable actions, and much more. As Kirsh puts it, "we adapt the environment instead of ourselves." All these manipulations use spatial location as the signifier.

Space is a very powerful tool, as illustrated by figure 8.3. Count the circles simply by looking at them: don't use your hands or a pointer to help. Difficult, isn't it? Counting itself is not a difficult task: the problem is keeping track of which items have been counted, and which haven't.

Now count the very same items shown in figure 8.4, again without using hands or other objects as aids: much easier, isn't it? These are the very same items, but organized into six clusters, all but one having precisely five circles. The organization makes it easy to count any one cluster because a small number can be counted at a glance, and by spreading the circles into six clusters distributed spatially, we can more easily move systematically from cluster to cluster, making it easy to remember which have been counted, which not.

The differences between the two figures change the task. In figure 8.3, the difficult part is the mental act of keeping track of which items have been counted while simultaneously charting a course through all the dots so that each is counted precisely once. In figure 8.4, as soon as one realizes that the dots are organized in

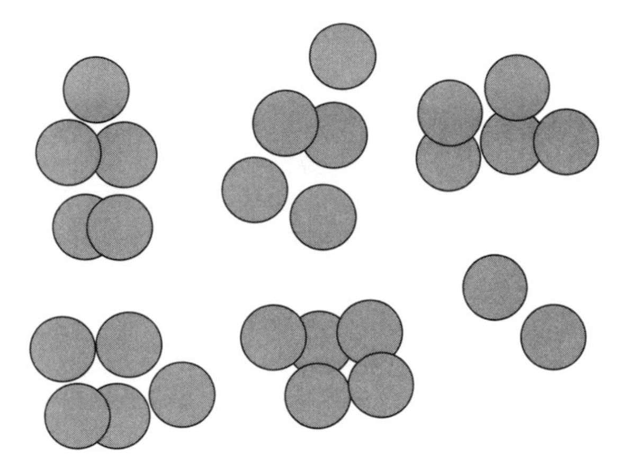

Figure 8.4
Cognitively congenial counting. Note how much easier this counting task is than that of figure 8.3. (The number of objects in figure 8.3 and figure 8.4 is the same.) Drawn following the description of the task in Kirsh 1995.

groups of five, it is a simple matter to count the number of groups, and their spatial separation makes keeping track easy. If these were physical disks in front of you, the problem could be solved by moving each disk into a "finished" pile as it was counted, very much like Kirsh's cooks separated washed from unwashed items.

All congenial methods work in this way: change the task into one that fits human cognitive structures and the difficulty is reduced. Much of the power of diagrams and pictures lies in the match between pictorial representations and the human perceptual system. Managing the environment not only brings structure to our tasks, it has important social benefits by making it easier to explain the tasks, easier for others to help.

Automation

Automation eliminates the need for doing a task. Most of our modern technologies get simpler with time through the increased use of automation. Thermostats maintain home temperature, often different for daytime than night, setting themselves differently when people are present than when they are absent. Microwave ovens and refrigerators are controlled by microprocessors. The electronic messaging systems perform very complex routing, a complex translation of human-readable addresses and names into precise, machine-understandable forms. Modern aviation would not function without considerable automation, nor would modern manufacturing plants and warehouse distribution systems. Automation adds an underlying complexity to the technology, but it simplifies the actions from the point of view of the people doing the work.

Automation only simplifies as long as it functions. When automation fails, it can make the task even more complicated than it would have been had there been no automation at all. Similarly, partial automation can be more problematic than either full automation or no automation, because switching between the automated and nonautomated states can add confusion and complication. I discuss these issues at length in my book *The Design of Future Things*. The important point for the treatment of complexity is that automation is perhaps the most effective simplifying strategy of all, as long as the function is completely automated by a system that is robust and reliable.

Beneficial Manipulation: Forcing Functions

"Forcing functions" are constraints intended to prevent unwanted actions. Forcing functions simplify tasks because no understanding is required: the function forces the intended behavior. It is only if one wants to perform the prohibited action that some understanding must be brought to bear.

Go back and look at the gate that blocks access to the stairs in figure 8.1b. Why is it blocked? In many places, it is not legal to have a staircase continue down from a higher floor of a building beyond the ground floor and into the basement area. The reason is that when there is fire, people escaping down the stairways might continue beyond the ground floor into the basement, where they could be trapped. The solution is to use a forcing function, preventing a mindless dash past the ground floor to the basement in emergencies, but permitting people who need to go to the basement to get there. This is done by blocking the stairway from the ground floor down, either with a door or gate, as in figure 8.1b, or by having the stairs end at the ground floor. The stairs leading to the underground floors are located elsewhere.

Sometimes behavior can be manipulated by making possible actions invisible—removing all signifiers. In walking through a Disney theme park, I was surprised when my host, a Disney executive, entered a nondescript alley, made a few turns, and led me to the backstage area. There were no doors, gates, or guards. The only thing that prevented park visitors from doing the same was the invisibility of the path.

Sometimes behavior can be manipulated by making it appear to be impossible, or at least difficult or dangerous. One way to do this is by using deliberately misleading signifiers: call these "negative signifiers." Examples are broken bottles or other sharp objects embedded at the top of a fence or wall, intended to prevent people from climbing over. Barbed wire is a negative signifier. Some parks use vertical pipes blocking the road, signifying that automobiles cannot continue. The park staff, however, knows that the "pipes" are actually flexible rubber tubes, so vehicles can ignore the apparent negative signifier and drive right over them.

Forcing functions play a role in the operating controls of many complex systems, preventing some operations from being performed until all the necessary prerequisites have been met or some safety precaution has been applied. In automobiles, it is not possible to start the car without depressing the brake, a clear safety function. In other cases, a critical set of controls is not made available until all the preceding interlocks have been removed. Interlock? That's a forcing function, preventing an action. Thus, the home microwave automatically turns off when its door is opened, preventing accidental radiation. Forcing functions are valuable aids to safety.

Nudges and Defaults

Forcing functions are valuable, but they are often too strong for the purpose. Not everything needs to be forced. Sometimes, all that is desired is a gentle nudge. Richard Thaler and Cass Sunstein, an economist and a lawyer, respectively, have devised a philosophy of beneficial manipulation that they call "nudging." Thaler and Sunstein looked at situations where people failed to behave in their own best interests, trying to understand why this is so. This probably applies to many of the readers of this book. Do you eat properly, exercise on a regular basis, save the proper amount of money for your retirement, and avoid excessive purchases on credit? Most people agree that all these behaviors would be beneficial for themselves, but most of us fail to do all of them appropriately. Why? That is what Thaler and Sunstein's book *Nudge* is all about.

Thaler and Sunstein show that designers have many subtle tools that can be used to control behavior. One subtle way of nudging is through clever placement of items in a list, or even in the foods displayed in a cafeteria, with healthy foods near the beginning and easily reached, unhealthy desserts and temptations at the end of the line, behind other things so they are not so easily reached. When people are choosing from a list of items, the first few items are most likely to be selected. During elections, the top name on the ballot has an advantage. Election officials often go to great lengths to randomize the placement of names to minimize the impact of this bias. With electronic displays on voting machines, each voter could see a different ordering of names, minimizing the subtle impact of order on the election result.

The word "default" refers to an action that takes place automatically, unless someone chooses otherwise. Defaults are subtle nudges to accept the default action, in part because they are so automatic and invisible. The withholding of income taxes from a paycheck is done by default. And when a person is initially hired for a job, all sorts of activities take place automatically following default conditions.

In the United States, employees of many firms were allowed to choose ways of investing part of their salary for use during retirement. The workers could choose to defer a considerable amount of their salary, tax free, to be put into an investment account, often one of their choosing, until retirement time. Sometimes the employer would provide some matching funds, adding to the amount being invested. This has been considered to be a good thing for all involved.

Despite the apparent virtues of the system, surprisingly few employees took advantage of it. Why? Because it required effort to think through the options and to make the decision. Generally, the option was only offered visibly once, at the time of initial employment, when many other choices were being demanded of the employee. The default was not to invest money.

Thaler and Sunstein regard defaults as one of the most powerful tools for manipulating behavior. In the case of saving for retirement, suppose the default was that a portion of the paycheck would automatically be invested each month in an investment account. Both methods of offering the investment opportunities are logically equivalent. In the first, if the employee did nothing, no investment would be made. In the other, if the employee does nothing, the investment would be made. These alternatives are called “opt in” versus “opt out.” It should come as no surprise that “opt out” has far more people taking part than the alternative. The United States Congress has mandated an “opt out” plan for voluntary pension contributions. They are just as voluntary as before, except now it happens unless you do something to stop it.

The use of defaults is an effective way to simplify interactions with the complex world in which we live. Defaults can’t be escaped, because any time a choice has to be made, the very way the alternatives are displayed presents some default actions. Even a refusal to make a choice is itself a choice. Although logically, there should be no difference between the two alternatives of opt in or opt out, logic and behavior are very different things. Defaults are powerful design tools, but they have to be used with care, both by the designer and by the person being confronted with them. To follow a default is to allow someone else to make the decision for you. This does indeed simplify decision making, but it is desirable only if you agree with the choice.

Learning Aids

The traditional way to explain the workings of a product is through an instruction manual. Most of them, however, are of little value: people don’t even read them. One reason we don’t read manuals is lack of motivation. Who wants to read a dull, dreary manual? Why not just get started? When people first use a new product or service, they have some goal they need to accomplish. They want to get to that goal, and reading a manual seems like a diversion.

Most people want “just-in-time” learning. People learn best when they have a need to learn. Although logically, instruction should come first, before the need arises (the theory being that in this way you will know what to do), before there is a need there is little interest to learn.

Many manuals try to list all the features of a product, sometimes in alphabetical order, describing what each control or feature does. This violates the principles of both motivation and just-in-time learning. The best explanations are to explain usage in context, by showing how particular tasks can be done. Instruction needs to be focused on the task to be accomplished. Yes, there is a need for a thorough description of features, but this is best put into an appendix, where it can be referred to when needed: it should never be the primary learning tool.

People learn by doing. Telling people what to do is not nearly as effective as coaching them as they do it. Granted, this is not very practical for most products or services, but an excellent replacement is short video demonstrations (with the emphasis on "short"). Videos can show the operations in context as concrete actions rather than as abstract descriptions, making them easy to understand. They should be brief and to the point. A ten- to thirty-second video is sufficient to demonstrate many operations. But the video should be a real demonstration of a task, not someone trying to sell the product or showing feature-by-feature all the things that can be done: these kinds of videos are counterproductive.

Manuals should be reserved for quick, efficient instructional material: short demonstrations, tutorials, and at the end, for those who need more advanced knowledge, a complete description of the purposes of all the features and options, with as many illustrations as possible. If it is necessary for the company to include legal cautions and other material, these should be placed elsewhere, so that they do not intrude upon the pleasant experience of the product. The manual is often considered an expensive add-on rather than an essential part of the product, and so it is left for last, done in a rush, delivered electronically to save money even though this makes it difficult to access and use. The people who write these manuals understand these issues well, but they are powerless to change the situation.

Even better than a good manual, however, is a system that doesn't need a manual. The best people to help design a product so that it does not need a manual are the technical writers who today write the manuals. They know the difficulties people face. They understand how difficult it is to explain the products. They can help design products that are easy to explain, if not self-explanatory.

The company should recognize that the best product is the one with the best experience. Why ruin an excellent experience with a manual that emphasizes possible dangers and legal cautions? Why destroy the experience with dull, dreary feature lists when it would be much more effective to show how to accomplish all the wonderful things that the product promises? Make manuals concise, productive, and essential parts of the experience.

Rules for the Rest of Us: Coping with Complexity

Just as designers must do their jobs to make products and services understandable, we must do ours, spending the time to understand and master them. No matter how well designed some things are, no matter how good the conceptual model, the feedback, the structure, and modularization, complex activities must still be mastered, sometimes requiring hours, days, or months of study and practice. This is the way it is in our complex world.

After the designers have fulfilled their end of the bargain, it is the turn of us who make use of the systems. The way for us to deal with complexity is primarily one of acceptance: accept that complex things need time and effort to be mastered, and half the battle is won. But if rules are wanted, they boil down to a simple set of suggestions.

Acceptance

Relax. Realize that life is complex. In other words, everyone has to learn to understand and use complex systems. You can learn it too. Yes, it may take time; but all the other skills you know took time to learn as well. Complexity is here to stay. The frame of mind is essential: learn to accept complexity, but also learn to conquer it. Complex things become simple once they are dealt with properly, once they are divided into small parts, each of which is relatively easy to master, once it is understood, and once the cues built into the system are discovered and used. The first step toward conquering complexity is acceptance.

Divide and Conquer

Divide the task into small, understandable modules. Learn one module at a time. Then, as each module is learned it provides a feeling of accomplishment that helps motivate the learning of the next.

Just-in-time Learning

Don't try to learn everything at once: learn only what is needed to do the task that interests you. Then slowly add other tasks. Slowly pick up the advanced features. Learn when it is needed.

Understand, Don't Memorize

Try to develop a conceptual model of the technology: what is it really doing? How does it work? If you can learn this, then quite often the operations will seem sensible, and when that happens, they become learnable. Unfortunately, many technologies seem to go out of their way to make this understanding difficult to acquire. Do your best to avoid these.

Watch Other People

Watch other people using the technology: see what they do and how they do it. Don't hesitate to ask for help or, more importantly, why they did what they did. "I saw that you did this," you might say to someone, "what were you doing?" Most people will be happy to help. This is how even experts learn little secrets that they were unaware of. It is how children learn even basic operations of life: watching their parents and mimicking their behavior. It is a natural, effective way of learning. Do it deliberately. Tell people what you are doing: "I'm just learning to use this, would it be OK to watch you for a while?" Announcing your intention avoids potential embarrassment when they wonder why you are watching, or worry that you might see some confidential information. It also triggers them to explain and be helpful.

Use Knowledge in the World: Signifiers, Affordances, and Constraints

Just as you might follow the trail left by others as you walk across a wooded area or a snow-covered city, search for the trails of others in using technology. Do what they have done: it is a great way to get started. Look for signifiers, whether natural physical ones, such as the trails left behind by people's actions; social ones, such as the activities and presence of people; or designed ones deliberately placed by the designer to aid you, but only if you attend to them. Discover the affordances: be creative in figuring out unusual or novel ways to do things. And exploit the power of constraints as guides for what you can and cannot do, should or should not do.

Use Knowledge in the World: Make Signs, Labels, and Markers
Throughout this book we have seen examples of the use of external markers, signs, painted lines, and even sticky paper dots and labels placed wherever needed. Take the initiative: whenever you have difficulty or confusion doing something, take a moment to reflect on which steps were most confusing and go back and annotate them. Use nail polish or notes, paint or marking pen. Don't be afraid to add what you need. The markers can be tasteful and attractive or ugly and intrusive: what matters is whether they help you complete your activities. It doesn't matter what you do: what matters is that you do something.

Use Knowledge in the World: Lists
One of the most powerful tools for taming technology is the list. It is also one of the least well understood, most maligned tools. Most of us make up lists of reminders: things we need to accomplish, items to buy at the grocery store. Lists provide a valuable supplement to our memories by putting the steps to be done or the action to be taken into physical reminders in the world.

One specialized list is called the checklist. With a grocery list, the items can be purchased in any order, and because they are seldom critical for safety, if an item is skipped or even if one is repeated, the resulting error is not serious. Checklists are often employed in safety-critical areas such as medicine, industry, or aviation, where the items are usually listed in sequence, where each item has to be done and checked off (often by a coworker) before being allowed to proceed to the next item. Checklists are especially important for people who must accomplish complex procedures, often while doing other activities and when subject to frequent interruptions.

Despite their importance and proven value, lists are not universally used. Why? In part, many people feel that to use a list is to question their ability. This is a special problem with safety-critical jobs performed by experts. After all, those people are experts at their jobs; why would they need reminders?

Human memory is fallible. Even when people are doing well-learned tasks, interruptions or unexpected difficulties can derail the procedure. Experience has shown that when the interruption or difficulty has been dealt with, people often have trouble remembering precisely the point at which the task should be resumed. Checklists solve this problem by containing an explicit list of what has been done, what remains to be done.

In aviation, where safe flying requires that a large number of items need to be checked, pilots and mechanics resisted checklists for years, believing that they were quite knowledgeable about their tasks and that checklists were demeaning, that they questioned their abilities. Meanwhile, many accidents were traced to accidental skipping of steps or setting values. Over a period of decades, checklists were slowly introduced into all commercial flight operations, used by both pilots and ground crew: they have proven to be invaluable.

Today, the use of checklists is common in commercial aviation. The pilots review the checklist together, one person reading the list aloud, the other checking the status or doing the operations called for. The accident rate has diminished.

In other disciplines, checklists are still being resisted. Medicine is one example. Practitioners pride themselves on their skills and specialized knowledge and are disdainful of attempts to standardize their work or—heavens—to require checklists. Numerous studies of medicine demonstrate the ability of medical checklists to reduce accidents, injury, and deaths. There are even best-selling, popular books on the topic. But checklists in medicine are still resisted. “Sure, those other doctors need checklists,” I have heard physicians say, “but not me. I know what I am doing.” They argue against standardization, reminding us that each patient is different, so that no single standardized list could possibly apply. And meanwhile, patients suffer.

Checklists do have limitations. Paper-based checklists do not readily allow the ordering of items to be changed. Sometimes it is not possible to do the steps in the order in which they appear on the checklist, and then the problem is, once an item has been skipped, how can the system ensure that it is remembered later? Electronic checklists solve this problem by keeping a list of items skipped and not done and then bringing them back at the end of the list.

A second limitation has to do with the election of items and procedures to put on the checklist. I have heard physicians explain that they will not use checklists because the items are wrong, or perhaps not always appropriate or even possible. These criticisms might be valid, but the criticism should be directed to the process by which the content is determined, not to the principle of the checklist itself. Checklists should always be scrutinized for relevance and accuracy. They should always be undergoing improvement. Checklists are like any other product. They must be designed with care, preferably using the standard human-centered design techniques: observational studies, prototypes, and continual refinement, using feedback from test sessions. None of this should detract from the importance of checklists and of reminders.

Lists have been proven to work. They do need to be written with care. They can continually be studied, analyzed, and refined. But make lists and use them. They are not signs of weakness: they are signs of strength, powerful tools to help us do our tasks better, with more confidence and fewer errors.

Complexity can be managed, but to do so, we all must play our part.

9
The Challenge

Complexity can be rewarding, but it is also a challenge. Complex activities, events, and objects can be deep and satisfying. Complexity provides for multiple experiences and opportunities for engagement. It is to be relished and sought after. But complexity by itself is not a virtue: ill-structured, ill-advised complexity can be confusing and frustrating. The challenge for the designer is to provide well-structured, cohesive experiences, where the complexity can reveal its desirable face, not its ill-tempered, mystifying one. The challenge for us is to take the time and effort to learn the structure and the power of the design. Even the most complex things are simple to those who have mastered the structure, understood its operations, and have a cohesive internal understanding—a good conceptual model. Simplicity is in the mind. The perception of simplicity requires the joint efforts of those who design and those who use.

Even though many things are complex out of necessity, not everything needs to be complex. Many otherwise simple things are overdesigned, far too complicated. Why is so much of our modern technology badly designed, disfigured with excessive features?

Why does the disease of featuritis strike so often despite the existence of remedies and preventative measures? Why are so many things needlessly complex, needlessly complicated?

There is a market for simple, easy-to-use devices. Take cell phones. Many people would like a mobile telephone that is simply a telephone. Sure, it should be able to recall the last few callers and to store telephone numbers, but it doesn't also have to be a music player, camera, navigation system, and all those other things. Just a phone, please. Some manufacturers have tried to deliver these less complicated products, but they are often thwarted by two major influences in the complex chain of distribution and sales: salespeople and reviewers.

The Bias of the Sales Force

In my classes, I ask students to examine these questions. Katherine Duff, one of my resourceful MBA students, interviewed a designer from a major design firm that was designing a new phone for one of the country's largest manufacturers of cell phones. Here is the story she emailed me, edited to delete the names:

> I spoke with (a designer) from X . . . and he told me a story about how he had designed a cell phone for Y. It was targeted at consumers older than 50 and in designing it, they had done everything we are taught in this class. Designers had gone out and watched people use their phones, they had observed problems, they had designed prototypes and they tested them with the target audience. In the end, they found that the consumer loved the finished

product which was slightly larger than typical cell phones, only had three features (calling, a contact list and an alarm clock) and had large buttons. However, the phone was a huge flop because they couldn't get salespeople to sell the phones. They weren't cool. They were too simple. They didn't have a CAMERA. And because of that, the majority of the target audience never even saw them.

The Gap between Designer and Customer

There are many steps between the designer and the customer. In this case, the designer works for a design firm and the manufacturer is the client. The companies that make cell phones usually don't sell them directly to the people who use them: they sell them to the cell-phone companies, the service providers, who in turn have their own stores where they sell phones and telephone services to customers. So the designer's client is the manufacturer and the manufacturer's client is the service provider. The service provider's client is their stores. And the store's clients are the people who use the phone. Moreover, as I just discussed, the salesperson who actually sells the item may not have the customer's best interest in mind.

The separation between the manufacturer and the person who actually uses the product is true in many industries. Manufacturers of home appliances such as stoves, refrigerators, and washing machines sell to distributors, who sell to stores. Quite often the purchaser of these items is the contractor or developer who builds homes and apartments and sells them equipped with kitchen appliances. The home dweller has little or no say in the choice of appliance. When the people who will use a device purchase it from a store, their purchase decisions are often guided by salespeople who earn their living through commissions, a payment for each sale, often based on its profitability. As a result, they are often biased to recommend higher-costing items. Sometimes salespeople are given special commission rates for particular items, deliberately biasing the recommendations. There are many factors involved in the purchase decision: how well the product fits the true needs of the purchaser can easily get lost.

In stores that sell modern high technology, the salespeople often take great pride in their ability to demonstrate all the many features. They can show off their superior knowledge by contrasting the strengths and weaknesses of competing products. Even the most well-intentioned salespeople can fall into the trap of falling in love with the features and capabilities of the most advanced products rather than focusing on the needs of the customer in front of them.

Think back to my quotation of the failure of the telephone that seemed to please so many people when it was tested during the design phase. I can imagine earnest salespeople puzzled: "Why would anyone buy this phone," they might ask themselves, "when it is so limited in its capabilities?" They would be honest and concerned, but because they themselves love the multiple features of the technology, they fail to recognize that not everyone wants all those features. The much-reduced phone does not fit their model of what people do with their phones.

It is possible to overcome this sales problem, but only if the manufacturer goes to great trouble to have the item demonstrate itself, without the "aid" of salespeople. This approach was taken by Kodak when they launched a line of digital cameras. By observing both customer use and the sales process they recognized that not only had digital cameras become complex and confusing to use for customers, but salespeople were also struggling to explain features and functions. Kodak launched a line of simple one-touch automatic cameras, along with a simple way of getting the pictures out of the camera and printed. Along with the product they provided an in-store sales system that considerably simplified the education and sales process. Customers could watch the demonstrations themselves, without a salesperson. Kodak EasyShare quickly became the fastest-selling digital camera in the category, despite having fewer features and selling at a price premium. Kodak's success is directly attributable to understanding the customer and designing an end-to-end experience that meets customer needs better than anyone else in the category.

The Bias of Reviewers

Another major bottleneck in the quest for simpler products comes from the people who review the products for newspapers, magazines, and Web sites. Technology reviewers are technology lovers. Camera reviewers complain if the camera does not offer the photographer a choice of modes and settings: both manual and automatic controls, both shutter speed and lens aperture preference choices, film-speed adjustments, white balance modes, and so on. Most of these choices are for the experts, and even experts get confused. The average user neither understands them nor wishes to. Automobile reviewers still judge the cars on acceleration time.

They test the cars on winding roads at high speed. They talk about over steer and under steer, handling under heavy acceleration or braking on demanding courses, even though most drivers will never experience these situations.

Reviewers know too much. Only a few try hard to consider what the average family needs, but this is difficult because they have become far too expert about the industries they cover. Consumer testing magazines try hard to address the needs of the average car buyer, but they still list and rate features.

Large stores have tens of thousands of different items on display. It costs these stores a lot of money to keep track of all that stock and keep it fresh. The huge amount of choice confuses shoppers. When stores try to simplify their offerings, reviewers complain. Thus, when one large hardware store cut back on the number of different brands it carried, arguing that people "are looking for simplification," a reporter who covered the story said that this strategy might "delight minimalists" but that "it also limits choice." Limits choice? Yes, that's the whole point.

Barry Schwartz, a psychologist at Swarthmore University who studies decision making, wrote a best-selling book on the topic; the title says it all: *The Paradox of Choice: Why More Is Less*. Or as the statement on the front cover of the book puts it: "Today's world offers us more choices, but ironically less satisfaction."

Social Interaction

The human brain and capabilities of infants change extremely slowly generation after generation. Technology, society, and culture change far more rapidly—technology changing the fastest, culture the slowest. Moreover, the technologies now in common use and those contemplated for the near future are vastly different from what was available even a few decades ago. In the twentieth century, our designs and principles applied to a single person using a single device. But now, in the twenty-first century, more and more of our technologies support groups who can be in continuous communication. Social computing is the norm, even when the people within a group are separated by distance or time, with some working collaboratively, some barely knowing one another. The wide range of possible interconnections has many potential benefits, but it poses complications as well. It is wonderful to keep in touch with people, to keep track of friends even after they have moved to other places. It is convenient to set up a workgroup quickly in order to solve some problem on the job or to do an assignment at school. But all these different groups soon produce complex network structures. The social groups overlap and sometimes conflict. Work can get intertwined with play, social interactions with serious business details. Keeping all these relationships straight can be difficult. The potential for continual interruption is huge. And then, as if life isn't complicated enough, we must also be vigilant for those who are deliberately out to stalk, steal, sabotage, or otherwise disrupt our activities, both in the private, personal part of our lives and in the world of commerce and business.

Designing for groups is not entirely the same as designing for an individual. The people within a group still have the same requirements they had when they were working alone. But now there is often a need to synchronize. If several people are collaborating on a project, it can be both stimulating and problematical to work on the very same part of the work simultaneously. Stimulating when the combined group produces work superior to what any individual could do. Problematical when there are conflicts, where people disagree about the exact form, content, or placement of ideas. A group has knowledge greater than the sum of its parts. Some of this knowledge is explicit, as when group members collaborate to solve a problem, or deliberately label or tag items for the benefit of others. Other knowledge is emergent, appearing out of the behavioral interaction, as, for example, when some people's actions create trails that others can follow through complex spaces. Another difference found in groups is the way people will form subgroups within the larger group. Sometimes one subgroup devises plots against others. Moreover, it is common for people within a group to form strong cohesive, supportive bonds within the group, but to distance themselves from other groups, sometimes exhibiting competitive behavior. Designing for social interaction and groups will be a major theme in the twenty-first century.

Why Simple Things Become Complex

Here is how complexity increases naturally. One day a company announces a music player. Soon the company makes it possible to show music videos, so now the music player can do two things: play music and play music videos. Soon, customers ask if they can play any video, for example ones they have taken themselves, ones sent to them by friends, ones found on networks, and, while we are at it, television shows and movies. These features get added, but after a while, customers wonder why they are restricted to music and video: why not regular photographs? This leads to the request for a camera to take the photos and video clips. In the modern world, none of this works unless the music player is wirelessly connected to networks so that things can be shared, but as long as the device is connected to a social network, why not allow sharing of messages, thoughts, and location? Over time, the simple music player has morphed into a monster. (And did I mention the ability to make phone calls, as long as it is on a network? What about reading books? Why not?) Note that although this story is fictional, there are several very successful products on the market that do all the things described in this fictional example, one of which started out as the little music player called iPod. The details of the story are fictional, but the end result is true.

Every time a new technology appears, people soon master its capabilities and ask for more. As our demands for services, functions, and features grow, it is inevitable that the complexity of our technology will increase. The most advanced use of sophisticated, complex technology for the everyday person can be found in the automobile. The complexity is reaching dangerous levels because the interaction necessary to control these systems is a distraction for the driver. Just when does a system become too complex, too complicated for safe use?

Although the science of complexity reduction is well understood, it is still not well practiced. Automobile cockpits have special requirements because their controls must work well with unskilled people, under stress, in limited time, where the device is usually not the major focus of attention. When a driver wishes to change the interior temperature, or change the radio station or music selection, this activity should not distract from the primary task of driving. This means that the instrument panel, already crowded with displays, navigational devices, controls for multiple passenger temperature settings, driving parameters, and entertainment must be able to be used by someone relatively unskilled at both the task at hand and even the primary task of driving. Even a short gap in attention to driving can be fatal. Well-documented studies of drivers show that accident rates increase sharply as the time that the gaze is off the road exceeds two seconds.

What's true in the automobile also applies in the home, although without the same time stress and safety concerns. Appliances are increasing in complexity. Washing machines and driers, dishwashers and microwave ovens, coffeemakers and refrigerators are all now available with complex menus, multiple choices, and microprocessors.

To make matters worse: the terrorists are coming! Yes, they really are. And not just terrorists, but crooks, thieves, mischief-makers, and the curious. All want access to our records and our lives. Our records are not secure, our means of identifying ourselves are laughably impoverished, and the distinctions between security, identification, and authentication are poorly understood, even by those implementing the systems that control our lives. Most efforts to make our lives more secure results in making them more complicated. There is often a trade-off between ease of use and security. Efforts to ensure perfect security (an impossibility) can create onerous security demands. But there is a paradox: the more thorough the demands of security, the less secure the result. Why? Because as we saw in chapter 3, when things become too complicated, people find ways to simplify them. When the demands of security get in the way of doing our jobs, we find ways around them. Chapter 3 discussed how we try to overcome the hurdles of security. We write passwords on paper, hiding them in insecure locations. We prop open doors, make copies of sensitive material—all because we are dedicated to getting the job done. Thus, even the honest workers can undermine the entire security apparatus. Here is a field in desperate need of sanity, one based not only on technical considerations but psychological and social ones as well. Yes, we want the benefits of technology. Yes, we want security and safety. Complexity is acceptable as long as it is intelligible and necessary. We want to avoid needless complications.

The Design Challenge

This book has addressed the issues of complexity and suggested several approaches to taming its impact. One solution to the problems of complexity and security is to add more layers of technology, for example using automation to eliminate many complicated activities, simplifying the demands upon people (chapter 8). Alas, this often solves one set of problems while introducing others. Remember Tesler's *Law of Irreducible Complexity* from chapter 2: when we add automation to simplify the demands upon people, we increase the complexity of the underlying technology. The more complex the underlying technology, the more opportunities for failure. Difficulties with automation have been documented in every domain that has been automated, yet each new domain fails to learn from others but must encounter the problems anew. When properly deployed, automation can reduce stress and workload, decrease errors and accidents. But if badly deployed it can do just the opposite: increase stress and workload, change the type of error and accident, often to a much larger extent than was possible in the nonautomated state. This is what happened in aviation and commercial industrial plants. It is now beginning to happen in medicine, in the home, and in automobiles.

Designers and engineers can learn from the introduction of technology to other domains. Of course it is far easier to say this than to do because each new domain has both a lot in common with others and a lot that is special and unique to itself. So designers face a difficult challenge. They must still deliver people the choices they want and need, reduce the complexity of security, yet maintain conceptual simplicity and safety from error, equipment failures, and deliberate attacks by thieves, terrorists, and pranksters. Automation can simplify things dramatically, but we must be cautious in its application to avoid the perils of inappropriate levels of automation.

The designer's role is difficult, with many challenges. In addition to all the functional, aesthetic, manufacturing, sustainability, and financial issues associated with the design of products, and the cultural, training, and motivational issues associated with services, the designer has to ensure that the end result communicates properly to those who must use it. This is the role of the conceptual model. It includes the perceptible signifiers that indicate the steps that have been followed, the current state, and what is yet to come. This is the place for just-in-time instruction, providing critical learning precisely when it is needed.

Living with Complexity: A Partnership

Living with technology is a partnership between the designers and us. Most of the time, we simply want to get something done, but instead, we may be required to master the intricacies of a complex system. But that is the way things work in the world. The technologies we use must match the complexity of the world: technological complexity is unavoidable.

Even the best of designs still require work on our part. Because human memory has its own quirks and limitations, it helps to put the required information in the world. We can do it for ourselves with signs and markers, notes and colored stickers. We can use lists, whether electronic or paper. We can use the reminders in electronic calendars, social networks, and other systems. We need to use help systems and manuals when we get stuck. We must do our part to learn the structure and underlying conceptual model of the technologies we use. We must take time to master the skills. With understanding, we can make complex systems simple and meaningful.

Living with technology will be an ongoing challenge, but a necessary one. Taming technology requires a partnership between the designers and those of us who use it. The designers must provide structure, effective communication, and a learnable, sociable interaction. We who use the results must be willing to take the time to learn the principles and underlying structure, to master the necessary skills. We are in a partnership with designers.

Notes

1 Living with Complexity: Why Complexity Is Necessary

The discussion of the infield fly rule comes from the "Official Rules: Definition of Terms" section of the Web site of Major League Baseball (mlb.com): <http://mlb.mlb.com/mlb/official_info/official_rules/definition_terms_2.jsp>.

Schoenberg's comment about musical notation is available in his reprinted works: see Schoenberg 1985.

I once argued (in Norman 1982) that it took 5,000 hours of practice to become an expert on any topic. More recently, Ericsson (2006) has argued the number should be 10,000 hours. Malcolm Gladwell (2008) has provided an eminently readable summary of all this work in his book *Outliers*.

2 Simplicity Is in the Mind

Chris Sugrue's artwork, *Delicate Boundaries*, can be seen on her Web site: <http://www.csugrue.com/delicateBoundaries/>.

The conceptual model of the water cycle diagram in figure 2.3 is the U.S. Geological Survey, Department of the Interior, drawing of the water cycle by John M. Evans. Downloaded on July 19, 2009, from <http://ga.water.usgs.gov/edu/watercycle.html>.

The quotation from the newspaper columnist H. L. Mencken is from Mencken 1917.

My introduction to the planishing hammer came from an email conversation with Hugh Meyers about simplicity. I am indebted to him for providing this example and helping me understand how complex systems can result from the combination of simple components.

The planishing hammer quotation is from Wikipedia: <http://en.wikipedia.org/wiki/Planishing>.

Larry Tesler's law of irreducible complexity has never been published, but it is discussed in an interview with him (Tesler and Saffer 2007). Bruce Tognazzini (Tog) has a nice discussion of the entire issue on his blog "Ask Tog": <http://www.asktog.com/columns/011complexity.html>.

Many philosophers over the centuries have postulated similar sayings about the nature of simplicity, all arguing that explanations should have as few postulates as possible. My source for William of Ockham is the *Encyclopaedia Britannica* (Ockham's razor 2010). My source for Einstein is Wikipedia, <http://en.wikipedia.org/wiki/Einstein>. I suspect that both Ockham and Einstein restated their beliefs multiple times, adding to the confusion about what they actually might have said. However, everyone agrees on the spirit of their words.

Several sections of this chapter are modified from my columns in the magazine *Interactions*, published by the ACM's professional association SIGCHI (Special Interest Group in Computer-Human Interaction).

3 How Simple Things Can Complicate Our Lives

Gwendolyn Galsworth's powerful ideas for adding structure to the workplace through visible lines and guides is available in her books: the title "Visual Workplace—Visual Thinking" comes from Galsworth 2005.

The quotation from Henri Aebischer about the BBC weather forecasts was part of an email conversation in January 2009. I am grateful to Henri for his many interactions with me on the subject of design and for permitting me to quote from his email.

The word "nudge" comes from Richard Thaler and Cass Sunstein's (2008) book *Nudge: Improving Decisions about Health, Wealth, and Happiness.* They show there that social forcing functions can be devised to help people make better economic decisions, yet without requiring them to do so.

4 Social Signifiers

The ways in which both implicit and explicit signifiers affect cognition has been very nicely described in an article by Hollan, Hutchins, and Kirsh (2000): "Distributed Cognition: A New Foundation for Human-Computer Interaction Research." They did not use the term "signifier," but did talk of the power of information in the world (external representations) to guide individual and group cognition—distributed cognition is the term Hutchins coined to describe this. The studies of "marks left behind by readers" and other related issues are found in Hill et al. 1992 and Hutchins 1995a,b.

The field of semiotics has a very rich and extensive literature. It has been applied to design most effectively by Clarisse de Souza (2005). Judith Donath's work on sociable media is especially relevant here: see Donath 2007 and especially Donath forthcoming.

The quotation "The field of semiotics is the study of signs" is from the Swiss linguist Ferdinand de Saussure, cited in *Encyclopaedia Britannica* (Semiotics 2010). The signaling behavior of the gazelle is described in Bliege Bird and Smith 2005 and Zahavi and Zahavi 1997.

An excellent discussion of how to manage the intricate rules of etiquette is given in a most enjoyable essay entitled "Rogov's Ramblings: Good Manners" at <http://www.stratsplace.com/rogov/good_manners.html>.

The etiquette quotation "place the napkin in your lap shortly after you sit down" comes from Van Der Leun 2005.

5 Design in Support of People

The source for the origin of the phrase “reticulating splines” is an interview with Will Wright, the developer of the SimCity and other Sims computer games: he inserted the phrase into his SimCity 2000 (SimCity 2000, 2008).

A summary of the early work comparing the interactions of people with people and people with machines was done by Byron Reeves, Cliff Nass, and Scott Brave at Stanford University (Nass and Brave 2005; Reeves and Nass 1996).

The instructions on how to thwart the will of people trying to find the easiest way across a landscaped space comes from Voice 2007. I met Carl Myhill at a conference in New Zealand, where he introduced me to desire lines and to the extension of the concept to many aspects of design. His work (Myhill 2004) inspired my treatment here.

The history of Paul Otlet’s *Traité de documentation* in 1934 and Vannevar Bush’s memex system come from several sources: Bush 1945; Vannevar Bush 2010; Wright 2003, 2008. I thank Jonathan Grudin for informing me of Otlet’s work.

6 Systems and Services

Some of the material in this chapter is adapted from studies performed with Kara Pernice of the Nielsen Norman group. We had originally thought that we would be writing a book entitled “Engaging the Customer,” but although a few draft chapters were produced, the effort came to naught.

James Teboul, professor of operations management at INSEAD, an international business school with campuses in France and Singapore, distinguishes the frontstage of a service from the backstage (Teboul 2006). Lynn Shostack’s proposal for service blueprints was published in the *Harvard Business Review* (Shostack 1984). The work by Susan Spraragen used in this chapter is reported in Spraragen 2010.

The patent discussion for the design of the Washington Mutual Bank comes from a newspaper article by Wu, "Bank Drops Drab" (Wu 2004), but the story is readily confirmed by searching the bank's public relations pages and patent files. The retreat by Chase Bank to traditional bank design, eliminating all of the former Washington Mutual Bank's innovations, is covered in Robin Sidel's article in the *Wall Street Journal*, "WaMu's Branches Lose the Smiles" (Sidel 2009) and by intelligent and sympathetic discussion in blogs by Jeffrey Pilcher at the Financial Brand, <http://thefinancialbrand.com/tag/jp-morgan-chase/>, and at the John Ryan blog, <http://www.johnryanblog.com/2009/04/bankers-1-wamu-0/>.

The quotation from The Köln International School of Design in Cologne, Germany, is from <http://kisd.de/subject_sd.html?&lang=en> (retrieved July 2008).

The story of the Fred Harvey restaurant chain in the late nineteenth and early twentieth centuries is wonderfully described by Brown and Hyer in the otherwise very academic *Journal of Operations Management* (Brown and Hyer 2007). There is a huge literature on services, although most of the work concentrates on efficiency rather than satisfaction of both customer and employee. Here are the ones I found most valuable: Brown and Hyer 2007; Glushko and Tabas 2007; Heskett et al. 1994; Heskett, Sasser, and Schlesinger 2003, 1997; Parasuraman, Zeithaml, and Malhotra 2005; Teboul 2006. Heskett and his colleagues report that being satisfied was not good enough—customers had to be very satisfied before they became loyal (Heskett et al. 1994).

The distinction between front- and backstages of services comes from Glushko and Tabas 2007; Teboul 2006. The study of the Acela Express train service is reported in Richardson and Oppenheimer. And the quotation about Ritz-Carlton comes from John Collins, a Human Resource Manager at the Ritz-Carlton, during a training session attended by the author of a *Harvard Business*

Review paper on the topic: "My Week as a Room Service Waiter at the Ritz" (Hemp 2002).

The Netflix story is documented on the official Netflix blog, <http://blog.netflix.com/search?q=profiles> (retrieved July 16, 2008) and on numerous Web sites. The customer's response was from "*geeksugar," <http://www.geeksugar.com/1749822> (retrieved July 16, 2008).

The impact of how companies recover from service deficiencies has been much studied. I suggest reading McCollough, Berry, and Yadav 2000. And finally, the impact of noise in the intensive care ward comes from Brandon, Ryan, and Barnes 2007.

7 The Design of Waits

Some of this chapter, especially the story of my experience on my flight between Chicago and San Francisco, is taken from my paper, "Designing Waits That Work," published in the *MIT Sloan Management Review* (Norman 2009b). The paper expands on David Maister's classic paper, "The Psychology of Waiting Lines" (Maister 1985).

Bob Sutton, Professor of Management Science and Engineering at Stanford University, discusses the memory of a Disney visit in his paper "Feelings about a Disneyland Visit" (Sutton 1992). The quotation that "Disney employees are taught to pay special attention to customers" is taken from one of Sutton's email messages to me. The importance of feeling in control is a continual theme in publications within the scientific literature on social psychology. See Ward and Barnes 2001 as an example. The importance of conceptual models is also a frequent topic of discussion in the design literature, but a good starting point is my *Design of Everyday Things* (Norman 2002).

The quote that it is better to improve the waiting experience by making it more attractive than by shortening it is from Pruyn and Smidts 1998.

The importance of a strong ending comes from many places, including my own work on the serial position curve of human memory many decades ago. The book *Human Memory* by Alan Baddeley (1998), although somewhat dated, is still one of the best books available on the psychology of memory. The research that explores how adding a slightly less unpleasant ending to an unpleasant event enhances one's perception was done by Daniel Kahneman. A good summary of the work occurs in his acceptance speech for the Nobel Prize (or to be more precise, his acceptance speech for the "Sveriges Riksbank Prize in Economic Sciences in Memory of Alfred Nobel," informally referred to as the Nobel Prize in Economics; Kahneman 2003a,b). Also see the discussion by Chase and Dasu about the importance of segmentation of pleasurable events and of getting unpleasant experiences out of the way early (Chase and Dasu 2001). The work on "rosy retrospection" comes from the work of Terence Mitchell and Leigh Thompson (Mitchell and Thompson 1994; Mitchell et al. 1997). My thanks to Leigh Thompson for her discussions on this topic at Northwestern University's Kellogg School of Management.

Elizabeth Loftus has done a considerable amount of research on the unreliability of eyewitnesses and the related unreliability of human memory, both of which are very easy to bias and thereby induce false memories. The false memories can be believed even more strongly than true ones. The reference in this chapter is to Braun-LaTour et al. 2004. Also see Braun and Loftus 1998; Sacchi, Agnoli, and Loftus 2007. Differential forgetting of positive and negative events was studied by Trope and Liberman (2003), but also see the discussion of the design implications by Chase and Dasu (2001).

The story about the way MacDonald's restaurant changed queuing behavior in Hong Kong comes from the *Encyclopaedia Britannica* article entitled "Cultural Globalization" (Watson 2008). The work on cognitive dissonance comes from Festinger's classic study (Festinger 1957).

8 Managing Complexity: A Partnership

People seldom read manuals, as numerous studies and personal observations can confirm. One study described it this way: "several studies have now been published (Leonard and Karnes, 2000) showing that only about 5% of vehicle owners report reading their owner's manual cover to cover. Rather, they consider the manual a reference source to be used when some specific information is needed." From Laughery and Wogalter 2008.

The quote from President Bush (the first president Bush) about the need to ensure that all Americans can program their VCRs is discussed in Gerstenzang's (1990) article on the president; but I also heard Bush's talk when he made the statement.

Richard Thaler and Cass Sunstein's (2008) book is *Nudge: Improving Decisions about Health, Wealth, and Happiness*. Leonhardt (2007, "Technology Eases the Path to Higher Tolls") discusses how these principles make the payment invisible and therefore painless, making it easy for the authorities to keep raising the toll amount.

David Kirsh's work on managing the environment in order to bring structure to tasks is described in Hollan, Hutchins, and Kirsh 2000; Kirsh 1991,1995, 1996, 2003.

The two early proponents of human-centered design who argued for the importance of sites, modes, and trails were the Swiss researchers Jurg Nievergelt and J. Weydert, who made the title of their paper self-explanatory: "Sites, Modes, and Trails: Telling the User of an Interactive System Where He Is, What He Can Do, and How to Get to Places" (Nievergelt and Weydert 1987).

Checklists are extensively studied in various fields of human factors, ergonomics, and aviation safety, among others. Their use in these areas, but with a heavy emphasis on medicine, is covered very nicely in Gawande's popular book *The Checklist Manifesto* (Gawande 2009).

References

Baddeley, A. D. 1998. *Human Memory: Theory and Practice* (rev. ed.). Boston, Mass.: Allyn & Bacon.

Bliege Bird, R., and E. A. Smith. 2005. Signaling theory, strategic interaction, and symbolic capital. *Current Anthropology* 46 (2):221–248. <http://www.journals.uchicago.edu/doi/abs/10.1086/427115>.

Brandon, D., D. Ryan, and A. Barnes. 2007. Effect of environmental changes on noise in the NICU. *Neonatal Network* 6 (4):213–218.

Braun, K. A., and E. F. Loftus. 1998. Advertising's misinformation effect. *Applied Cognitive Psychology* 12:569–591. <https://webfiles.uci.edu/eloftus/BraunLoftusAdvertisingMisinfoACP98.pdf>.

Braun-LaTour, K. A., M. S. LaTour, J. E. Pickrell, and E. F. Loftus. 2004. How and when advertising can influence memory for consumer experience. *Journal of Advertising* 33 (4):7–25. <https://webfiles.uci.edu/eloftus/BraunLaTourPickLoftusJofAd04.pdf>.

Brown, K. A., and N. L. Hyer. 2007. Archeological benchmarking: Fred Harvey and the service profit chain, circa 1876. *Journal of Operations Management* 25 (2):284–299. <http://www.sciencedirect.com/science/article/B6VB7-4KKFPFX-1/1/243b7888dd026e69c5ff19d2aa7ecd40>.

Bush, V. 1945. As We May Think. *Atlantic Monthly* (July):101–108. <http://www.theatlantic.com/doc/194507/bush>.

Chase, R. B., and S. Dasu. 2001. Want to perfect your company's service? Use behavioral science. *Harvard Business Review* 79 (6):78–84.

de Souza, C. S. 2005. *The Semiotic Engineering of Human Computer Interaction*. Cambridge, Mass.: MIT Press.

Donath, J. 2007. Virtually trustworthy. *Science* 317:53–54. <http://smg.media.mit.edu/Papers/Donath/VirtuallyTrustworthy.pdf>.

Donath, J. Forthcoming. *Designing Sociable Media*. Cambridge, Mass.: MIT Press. <http://smg.media.mit.edu/people/Judith/signalsTruthDesign.html> (chapter abstracts).

Ericsson, K. 2006. The influence of experience and deliberate practice on the development of superior expert performance. In *The Cambridge Handbook of Expertise and Expert Performance*, ed. K. A. Ericsson, N. Charness, and P. J. Feltovich, 683–703. Cambridge: Cambridge University Press.

Festinger, L. 1957. *A Theory of Cognitive Dissonance*. Stanford, Calif.: Stanford University Press.

Galsworth, G. D., ed. 2005. *Visual Workplace, Visual Thinking: Creating Enterprise Excellence through the Technologies of the Visual Workplace*. Portland, Ore.: Visual-Lean Enterprise Press.

Gawande, A. 2009. *The Checklist Manifesto: How to Get Things Right*. New York: Metropolitan Books.

Gerstenzang, J. 1990. The President: Bush's humor. *Los Angeles Times*, May 7, p. A5 (San Diego edition).

Gladwell, M. 2008. *Outliers: The Story of Success*. New York: Little, Brown.

Glushko, R. J., and L. Tabas. 2007. Bridging the "front stage" and "back stage" in service system design. Berkeley: School of Information, University of California, Berkeley. <http://repositories.cdlib.org/ischool/2007-013>.

Hemp, P. 2002. My week as a room-service waiter at the Ritz. *Harvard Business Review* 80 (6):50–62. <http://ged.insead.edu/fichiersti/hbr2002/306040.pdf>.

Heskett, J. L., T. O. Jones, G. W. Loveman, W. E. Sasser, and L. A. Schlesinger. 1994. Putting the service-profit chain to work. *Harvard Business Review* 72 (2):164–174.

Heskett, J. L., W. E. Sasser, and L. A. Schlesinger. 1997. *The Service Profit Chain: How Leading Companies Link Profit and Growth to Loyalty, Satisfaction, and Value*. New York: Free Press.

Heskett, J. L., W. E. Sasser, Jr., and L. A. Schlesinger. 2003. *The Value Profit Chain: Treat Employees Like Customers and Customers Like Employees*. New York: The Free Press.

Hill, W., J. D. Hollan, D. Wroblewski, and T. McCandless. 1992. Edit wear and read wear: Text and hypertext. In *Proceedings of the 1992 ACM Conference on Human Factors in Computing Systems (CHI'92)*. New York: ACM Press.

Hollan, J. D., E. Hutchins, and D. Kirsh. 2000. Distributed cognition: A new foundation for human-computer interaction research. In *ACM Transactions on Human-Computer Interaction: Special Issue on Human-Computer Interaction in the New Millennium* 7(2), 174–196. <http://hci.ucsd.edu/lab/hci_papers/JH1999-2.pdf>.

Hutchins, E. 1995a. *Cognition in the Wild*. Cambridge, Mass.: MIT Press.

Hutchins, E. 1995b. How the cockpit remembers its speeds. *Cognitive Science* 19:265–288. <http://hci.ucsd.edu/lab/hci_papers/EH1995-3.pdf>.

Kahneman, D. 2003a. A perspective on judgment and choice: Mapping bounded rationality. *American Psychologist* 58 (9):697–720.

Kahneman, D. 2003b. Maps of bounded rationality: A perspective on intuitive judgment and choice. In *Les Prix Nobel 2002*, ed. T. Frangsmyr Stockholm, Sweden: Almquist & Wiksell International. <http://nobelprize.org/nobel_prizes/economics/laureates/2002/kahnemann-lecture.pdf>.

Kirsh, D. 1991. When is information explicitly represented? In *Information, Language, and Cognition*, ed. P. P. Hanson, 340–365. New York: Oxford University Press.

Kirsh, D. 1995. The intelligent use of space. *Artificial Intelligence* 73 (1–2):31–68.

Kirsh, D. 1996. Adapting the environment instead of oneself. *Adaptive Behavior* 4 (3–4):415–452.

Kirsh, D. 2003. Implicit and explicit representation. In *Encyclopedia of Cognitive Science*, ed. L. Nadel, 478–481. London: Nature Publishing Group. <http://adrenaline.ucsd.edu/kirsh/articles/implicit_explicit/implicit_explicit.pdf>.

Laughery, K. R., and M. S. Wogalter. 2008. On the symbiotic relationship between warnings research and Forensics. *Human Factors* 50 (3):529–533.

Leonard, S. D., and E. W. Karnes. 2000. Compatibility of safety and comfort in vehicles. Paper presented at the Proceedings of the IEA 2000/HFES 2000 Congress.

Leonhardt, D. 2007. Technology eases the tide to higher tolls. *New York Times*, July 4.

Maister, D. 1985. The psychology of waiting lines. In *The Service Encounter: Managing Employee/Customer Interaction in Service Businesses*, ed. J. A. Czepiel, M. R. Solomon, and C. F. Surprenant. Lexington, Mass.: D. C. Heath and Company, Lexington Books. <http://davidmaister.com/articles/5/52>.

McCollough, M. A., L. L. Berry, and M. S. Yadav. 2000. An empirical investigation of customer satisfaction after service failure and recovery. *Journal of Service Research* 3 (2):121–137.

Mencken, H. L. 1917. *The Divine Afflatus* in *New York Evening Mail* (November 16, 1917); later published in *Prejudices: Second Series*

(1920) and *A Mencken Chrestomathy* (1949). Retrieved from <http://en.wikiquote.org/wiki/H._L._Mencken> on May 17, 2008.

Mitchell, T., and L. Thompson. 1994. A theory of temporal adjustments of the evaluation of events: Rosy prospection and rosy retrospection. In *Advances in Managerial Cognition and Organizational Information-Processing* (vol. 5), ed. C. Stubbart, J. Porac, and J. Meindl, 85–114. Greenwich, Conn.: JAI Press.

Mitchell, T. R., L. Thompson, E. Peterson, and R. Cronk. 1997. Temporal adjustments in the evaluation of events: The "rosy view." *Journal of Experimental Social Psychology* 33 (4):421–448.

Myhill, C. 2004. Commercial success by looking for desire lines. Paper presented at the 6th Asia Pacific Computer-Human Interaction Conference (APCHI 2004). <http://www.litsl.com/personal/commercial_success_by_looking_for_desire_lines.pdf>.

Nass, C. I., and S. Brave. 2005. *Wired for Speech: How Voice Activates and Advances the Human-Computer Relationship*. Cambridge, Mass.: MIT Press.

Nievergelt, J., and J. Weydert. 1987. Sites, modes, and trails: Telling the user of an interactive system where he is, what he can do, and how to get to places. In *Readings in Human-Computer Interaction: A Multidisciplinary Approach*, ed. R. M. Baecker and W. Buxton, 438–441. San Francisco: Morgan Kaufmann.

Norman, D. A. 1982. *Learning and Memory*. New York: Freeman.

Norman, D. A. 2002. *The Design of Everyday Things*. New York: Basic Books. (Originally published as Norman, D. A. 1988. *The Psychology of Everyday Things*. New York: Basic Books.)

Norman, D. A. 2007. *The Design of Future Things*. New York: Basic Books.

Norman, D. A. 2009a. Compliance and tolerance. *Interaction* 16 (3):61–65.

Norman, D. A. 2009b. Designing waits that work. *MIT Sloan Management Review* 50 (4): 23–28.

Ockham's razor. 2010. Encyclopaedia Britannica. Retrieved from <http://www.britannica.com/EBchecked/topic/424706/Ockhams-razor> on January 29, 2010.

Parasuraman, A., V. A. Zeithaml, and A. Malhotra. 2005. E-S-QUAL: A multiple-item scale for assessing electronic service quality. *Journal of Service Research* 7 (3):213–233. <http://jsr.sagepub.com/cgi/content/abstract/7/3/213>.

Planishing. 2009. Retrieved from <http://en.wikipedia.org/wiki/Planishing> on January 15, 2010.

Pruyn, A., and A. Smidts. 1998. Effects of waiting on the satisfaction with the service: Beyond objective time measures. *International Journal of Research in Marketing* 15 (4):321–334.

Reeves, B., and C. I. Nass. 1996. *The Media Equation: How People Treat Computers, Television, and New Media Like Real People and Places*. Stanford, Calif.: CSLI Publications and New York: Cambridge University Press.

Richardson, B. J., and B. Oppenheimer. Acela. *@issue Journal* 7(2): 24–31. <http://www.cdf.org/issue_journal/acela.html>.

Sacchi, D. L. M., F. Agnoli, and E. F. Loftus. 2007. Changing history: Doctored photographs affect memory for past public events. *Applied Cognitive Psychology* 21:1005–1022. <https://webfiles.uci.edu/eloftus/Sacchi_Agnoli_Loftus_ACP07.pdf?uniq=je5vga>.

Schoenberg, A. 1985. A new twelve-tone notation. In *Style and Idea: Selected Writings of Arnold Schoenberg*, ed. A. Schoenberg and L. Stein (354–362). Berkeley: University of California Press. (Originally published in 1924.)

Schwartz, B. 2005. *The Paradox of Choice: Why More Is Less.* Hampshire: Palgrave Macmillan.

Semiotics. 2010. In *Encyclopaedia Britannica*. Retrieved February 25, 2010, from Encyclopaedia Britannica Online: <http://www.britannica.com/EBchecked/topic/534099/semiotics>.

Shostack, G. L. 1984. Designing services that deliver. *Harvard Business Review* 62 (1): 133–139.

Sidel, R. 2009. WaMu's branches lose the smiles. *Wall Street Journal*, April 7, p. C1, from <http://online.wsj.com/article/SB123906012127494969.html>.

SimCity 2000. 2008. Retrieved from <http://en.wikipedia.org/wiki/Simcity_2000>.

Spraragen, S. 2010. Practicing the best practice: Designing effective health care experiences. Paper presented at the Design & Emotion, 2010 Conference, Chicago.

Spraragen, S., and C. Chan. 2009. *IBM Service Design Workbook: Expressive Service Blueprinting: Setting the Stage for Positive Customer Experiences*. Distributed at workshop given at the Art and Science of Service V conference.

Steinbeck, J. 1952. *East of Eden*. New York: Viking Press. <http://www.scribd.com/doc/24313691/John-Steinbeck-East-of-Eden>.

Sutton, R. I. 1992. Feelings about a Disneyland visit: Photography and the reconstruction of bygone emotions. *Journal of Management Inquiry* 1 (4):278–287.

Teboul, J. 2006. *Service Is Front Stage: Positioning Services for Value Advantage*. Hampshire: Palgrave Macmillan.

Technology. 2008. Encyclopaedia Britannica Online. Retrieved from <http://www.britannica.com/EBchecked/topic/585418/technology>.

Tesler, L., and D. Saffer. 2007. Larry Tesler interview: The laws of interaction design. In *Designing for Interaction: Creating Smart Applications and Clever Devices*, ed. D. Saffer. Berkeley, Calif.: New Riders. Published in association with AIGA Design Press.

Thaler, R. H., and C. R. Sunstein. 2008. *Nudge: Improving Decisions about Health, Wealth, and Happiness*. New Haven, Conn.: Yale University Press.

Trope, Y., and N. Liberman. 2003. Temporal construal. *Psychological Review* 110 (3):403–421.

Van Der Leun, J. 2005. Please don't drink the fingerbowl. *O, The Oprah Magazine*, August. <http://www.oprah.com/omagazine/Please-Dont-Drink-the-Finger-Bowl>.

Vannevar Bush. 2010. In *Encyclopaedia Britannica*. Retrieved February 25, 2010, from Encyclopaedia Britannica Online: <http://www.britannica.com/EBchecked/topic/86116/Vannevar-Bush>.

Voice, P. 2007. Desire lines and their part in landscaping. *Landscape Juice*, May 13. Retrieved July 13, 2008, from <http://www.landscapejuice.com/2007/05/desire_lines_in.html>.

Ward, J. C., and J. W. Barnes. 2001. Control and affect: The influence of feeling in control of the retail environment on affect, involvement, attitude, and behavior. *Journal of Business Research* 54 (2):139–144.

Watson, J. L. 2008. Cultural globalization. Retrieved May 10, 2009, from Encyclopaedia Britannica Online: <http://www.britannica.com/EBchecked/topic/1357503/cultural-globalization>.

Whitehead, A. N. [1920] 1990. *The Concept of Nature*. Cambridge: Cambridge University Press.

Wright, A. 2003. Forgotten forefather: Paul Otlet. *Boxes and Arrows*. Retrieved July 13, 2008, from <http://www.boxesandarrows.com/view/forgotten_forefather_paul_otlet#comments>.

Wright, A. 2008. The web time forgot: The Mundaneum Museum honors the first concept of the World Wide Web. *New York Times*, June 17. Retrieved from <http://www.nytimes.com/2008/06/17/science/17mund.html?pagewanted=all>.

Wu, A. 2004. Bank drops drab: Washington Mutual wins patent for branch concept. *SDGate (San Francisco Chronicle)*, June 26. Retrieved from <http://www.sfgate.com/cgi-bin/article.cgi?f=/c/a/2004/06/26/BUGND7CI8A1.DTL>.

Zahavi, A., and A. Zahavi. 1997. *The Handicap Principle: A Missing Piece of Darwin's Puzzle*. New York: Oxford University Press.

Acknowledgments

Complexity has shadowed my life in multiple ways, one of which being the construction of this book. My thoughts about complexity have had a long genesis. Many decades ago I was a fierce opponent, arguing strenuously for simplification. But over time, I came to realize that the enemy was not complexity, it was confusion and the resulting incoherence. Moreover, the solution was not simplicity as it is usually measured, which means through few controls, displays, and features, but rather coherence and understanding. I first tried these thoughts out in my column in *Interactions Magazine*, the publication of the professional society that studies human–machine interaction (the ACH SIGCHI), starting in 2007 and continuing to the present. So my first acknowledgment must go to the magazine and its hardy band of editors since 2007 who have given me permission to test my heretical ideas within its pages: Richard Anderson and Jon Kolko.

I have benefited from many opportunities to teach and present lectures on the material, each opportunity providing valuable feedback. Many people have held detailed conversations, helping me understand my own message. And many people have aided in the final assembly of examples, gracefully allowing me to use their photographs and drawings. My long-term friend and debating partner (and in former years, long-term coauthor of many research papers) is Danny Bobrow, who is always able to penetrate my ideas and find their weaknesses. Jakob Nielsen, my partner in business, continually provides me with a continual insightful examination of the foibles of design.

Some have read various segments of early versions of the manuscript and provided valuable feedback, such as Felix Portnoy and Henri Aebischer. Their comments have been most helpful. Many people have sent me photographs and story items over the years, some of which made it into this book, and I am grateful to all of them. I thank Chris Sugrue for her art work in chapter 2, Iain Tate for his photograph of the "not an exit" door of chapter 3, Kevin Fox for his photograph of the desire lines on the UC Berkeley campus in chapter 5, and Susan Spraragen for redrawing her figures for me (chapters 6 and 7). Jeffrey Herman enthusiastically helped my discussion of the planishing hammer and silversmith's bench of chapter 2 by sending photographs of his workbench and tools, and then telephoning me to talk about silversmithing. Silversmithing, alas, is a dying art in the United States.

My colleagues at Northwestern University, especially Ed Colgate and Liz Gerber, have allowed me to teach in their program, and my many MMM students, the dual-degree program in design and operations that I have codirected, offering both an MBA and an engineering degree, have suffered through my stumbling attempts to explain the mysteries of the design process. These are the people who will shape the products of the future, so my stumbles have consequences. It helps to have a most supportive dean, Julio Ottino, a champion of design who believes all engineers should be whole-brain thinkers: left as well as right brain, analytical as well as holistic.

Sandy Dijkstra and her staff at the Sandra Dijkstra Literary Agency have stood patiently by while this book went through its many iterations, starting out as something called "Sociable Design," before it found its true home in complexity. The staff at MIT Press have helped, rejecting my first attempt as confused, complicated, wordy, and redundant, but nurturing the final manuscript. Thanks to Katie Helke who worked hard to get the necessary figures and permissions, Judy Feldmann who deleted and edited, and most of all, to Doug Sery my editor, for patient support.

And, of course, my wife Julie, who has long served as my editor and most severe critic, never afraid to tell me the truth about my tendency to ramble, to repeat, to change point of view repeatedly, confusingly, and sometimes even within a single sentence. And other sins of writing and thinking. Every author should have such a truthsayer.

Index